建筑 100 BUILDINGS
1900–2000

[美] THE NOW INSTITUTE 编著
张涵 樊敏 译

中国建筑工业出版社

著作权合同登记图字：01-2019-2204号
图书在版编目（CIP）数据

建筑100：1900-2000 / 现在研究所（The Now Institute）编著；张涵，樊敏译. -- 北京：中国建筑工业出版社，2019.5
书名原文：100 Buildings（1900-2000）：Every Studens Should Know
ISBN 978-7-112-23611-4

Ⅰ.①建… Ⅱ.①现…②张…③樊… Ⅲ.①建筑设计—作品集—世界—现代 Ⅳ.①TU206

中国版本图书馆CIP数据核字(2019)第069610号

100 Buildings by The Now Institute
All rights reserved. No part of this publication may be reproduced, stored in a retrieval system, or transmitted in any form or by any means, electronic, mechanical, photocopying, recording, or otherwise, without prior consent of the publisher.
© 2017 Morphosis
Text by Val K. Warke and forecoord by Thom Mayne
Translation copyright © 2019 China Architecture & Building Press
Originally published in English under the title *100 Buildings* in 2017.
Published by agreement with Rizzoli International Publications, New York through the Chinese Connection Agency, a division of The Yao Enterprises, LLC.

责任编辑：李　婧　戚琳琳
责任校对：王　瑞

建筑100 （1900-2000）
[美] The Now Institute　编著
张　涵　樊　敏　译
*
中国建筑工业出版社出版、发行（北京海淀三里河路9号）
各地新华书店、建筑书店经销
北京点击世代文化传媒有限公司制版
北京富诚彩色印刷有限公司印刷
*
开本：850×1168毫米　1/32　印张：9⅛　插页：1　字数：347千字
2019年7月第一版　2019年7月第一次印刷
定价：68.00元
ISBN 978-7-112-23611-4
（31932）

版权所有　翻印必究
如有印装质量问题，可寄本社退换
（邮政编码 100037）

目 录

前言 / 005
汤姆·梅恩

致谢 / 007

建筑 100 / 011

矩阵图 / 213

建筑师的选择 / 215

英汉对照表与注释 / 247

参考文献 / 259

图片来源与英文版参与人员 / 284

译后记 / 289

前　言

汤姆·梅恩（THOM MAYNE）

  在最近十年的建筑教学中，我发现学生对历史案例的认知水平正在逐步下降。泛泛而粗浅的案例学习导致他们失去了发起真正有意义的建筑讨论的机会。究其原因，也许是数字媒体的风行推波助澜，抑或是某些更加无形的文化变迁产生了影响。最近我参与的一次设计评审会就极富代表性。我们看到的一项设计作品明显类同于黑川纪章（Kisho Kurokawa）位于银座（Ginza）的中银舱体大楼，这对于现场评委而言是显而易见的，学生对此却全然不知。由于她既没有听说过黑川纪章，也不知道舱体大楼，答辩因此而草草收尾。假若这位学生学过相关的建筑案例，她本可以就这项作品与新陈代谢派之间的关联展开讨论，也能够将自己的毕业设计置于一个更加宽广的知识框架之内。同事们交流之后，发现这种事一点儿也不稀奇。毫无疑问，去认识和欣赏那些最重要、最富影响力的建筑作品对于建筑系学生和非专业人士都有益处，今后建筑的发展也将从中获得启示。

  对于广大学生和建筑爱好者而言，这本书可以扩展他们对20世纪那些丰富多彩的、影响力强大的建筑佳作的理解，是极好的参考资料。我邀请了一批曾广泛参与工程实践的建筑师作为评委参与本书的选题：请他们各自列出学生应当知晓的20世纪100个建筑案例。不出所料，这些风格迥异的设计家选择的建筑简直五花八门——事实上他们总共提出了数千个方案；我们将这些方案以被选频次排序，选出前100名编入了本书，同时还绘制了一张案例与建筑师的矩阵关系图。最终的选题名单能够代表评委们的普遍性选择，其中的每个作品都在20世纪建筑发展的进程中发挥了关键作用。

  我认为这些20世纪的建筑佳作具有不可估量的价值，我们应当积极学习并努力理解其中的深意。这一时期的建筑实践范围极广，皆体现出巨大的社会、文化与政治抱负。通过本书，我们将了解建筑先行者们对专业与公众产生的巨大影响，也将认识到他们所面临的机遇与取得的成就——这无疑会扩展和激发我们的创造力。

致　谢

回望自己的建筑旅程，我发现 20 世纪早期的建筑和其后的经典作品都对我产生了深远的影响。正是这些建筑塑造、激励、挑战和鼓舞了我的建筑实践。借由众人的奉献精神，我们才能够将数量庞大的基础资料汇总统合为一本图书。

我首先要感谢 Eui-Sung Yi。他是 The Now Institute[1]的主任，也是墨菲西斯事务所（The Morphosis Architects）的负责人。我的学术研究、教学工作和建筑实践都离不开他的鼎力支持，迄今为止我们的合作超过十五年了。作为一名敬业的教师和重要的领导者，他编纂并审查了本书中 100 个不容忽视的建筑案例。五年前我们便开始筹划本书，Eui-Sung Yi 锲而不舍的努力和他对建筑可以改良生活这一信念的坚持是本书得以实现的关键。

没有墨菲西斯事务所的团队支持，本书便无法实现。经营主管布兰登·韦林（Brandon Welling）的勤奋与冷静使我们能够专注于工作；妮科尔·迈耶（Nicole Meyer）与萨拉·莫斯利（Sarah Moseley）辛勤写作，不断重整、编辑文本内容；莉莉·巴赫希（Lily Bakhshi）负责本书的设计装帧，她卓越的设计天赋毋庸置疑。

本书的核心内容以 The Now Institute 团队成员的研究成果为基础，逐步发展得来。瑞安·多伊尔（Ryan Doyle）所做的协调工作极为重要。凯文·谢罗德（Kevin Sherrod）、约翰·保罗·萨尔希多（John Paul Salcido）、贝扎·帕克索伊（BeyzaPaksoy）、陈奕涛（Yitao Chen）与当韦（Way Tang）等人则各有贡献，支持了本书的编写。

我所邀请的各位建筑师皆慷慨地提供了自己的建筑名单，他们为建筑行业做出了巨大的贡献，其观点保证了本书内容的专业性与有效性。我们将差异巨大的建筑作品置于一堂，希望能够将不同的建筑智慧与创新力联合起来，引发持续的研究与思考。

必须要感谢我亲爱的朋友瓦尔·沃克（Val Warke），他慷慨地贡献出自己的时间，保证了这本书的完整性。最后感谢我的妻子布莱思（Blythe），她是我生活的坚实

1　The Now Institute 是加州大学洛杉矶分校艺术与建筑学院名下的教学研究机构。——译者注

基础,也是我思想的指路明灯。

——汤姆·梅恩

还要感谢

阿纳什·托斯通伊斯(Ana Tostōes),DOCOMOMO国际协会

休伯特-简·汉凯特(Hubert-Jan Henket),DOCOMOMO国际协会

加州大学洛杉矶分校建筑与城市设计系及其学生

南加州建筑学院及其学生

南加州大学建筑学院及其学生

Shareefa Abdulsalam	Sai Rojanapirom
Nick Bruni	Dunia Abu Shanab
Lori Choi	LuyanShen
Çağdaş Delen	Jihun Son
NiloufarGolkarihagh	Niketa Sondhi
Ran Israeli	DevikaTandon
Sara Jafarpour	Rizzie Walker
Barak Kazenelebogen	Crystal Wang
Grace Ko	Yake Wang
Pegah Koulaeian	Tessa Watson
Deborah Liu	Robin Williams
Elisabet Ollé	BaoCheng Yang
Rupal Rathi	Halina Zárate

关于 The Now Institute

The Now Institute 是加州大学洛杉矶分校建筑与城市设计系主持的一所城市规划研究中心,重点关注城市策略的调查及应用,以解决城市弹性、文化、可持续性与城市流动性方面的复杂问题。在普利兹克建筑奖得主、著名教授汤姆·梅恩及 Eui-Sung Yi 主任的带领下,The Now Institute 将学术研究与专业需求整合,足迹跨越了美国和世界各地的城市。包括洛杉矶(Los Angeles)、新奥尔良(New Orleans)、马德里(Madrid)、北京(Beijing)、太子港(Port-au-Prince)和海地角(Cap-Haïtien)。

在十多年的研究历程中,The Now Institute 与市民、商业领袖、开发商、建筑师、城市规划师、文化制作人、学生及广大公众展开广泛的合作,完成了一系列极具可操作性的城市规划方略。代表性案例包括洛杉矶基础设施可持续性发展战略,海地水资源、教育与基础设施建设解决方案,食品来源不平衡背景下的城市农业策略,以及亚洲人工岛的地缘文化问题研究。

关于国际现代建筑遗产保护理事会

国际现代建筑遗产保护理事会（International Committee for the Documentation and Conservation of Buildings,Sites and Neighborhoods of the Modern Movement）是一个非营利性组织，致力于记录与保护现代建筑运动遗留的房屋、基址与社区。主要工作包括以下方面：

·向公众、政府、专业人士和教育界介绍、推广现代建筑运动的重要意义。

·考察并推动现代主义建筑的测绘工作。

·抵抗针对重大现代主义建筑作品的破坏与毁坏性改造。

·为保护现代主义建筑遗产或现代主义建筑（再）利用募集资金。

为了达成以上目标，国际现代建筑遗产保护理事会希望能将其行动拓展到新领域，与活跃于现代建筑领域的机构、组织与非政府团体建立新的伙伴关系，推行并发展国际现代主义建筑注册制度，并且扩大其在研究、文献与教育领域的活动范围。

The Now Institute 致读者的一封信

将一个设计方案变成一栋实实在在的建筑物是一件极具挑战的事情，其过程往往非常艰难。只有创意十足的概念可不够，还必须整合技术、政治、经济条件和其他的社会约束，做出综合考虑。因此，本书100个建筑实例中的每个作品都代表了其建筑师所取得的谦卑而鼓舞人心的成就。结合建筑所处的时代，这些作品绝对值得我们尊敬。

众多成就斐然的建筑师群力群策，共同为本书列出了一份伟大建筑名录。我们请每个人列出100个对青年建筑师的教育而言意义重大的建筑，所选作品必须是1900年至2000年建造的，以展示在20世纪变化的社会背景下现代建筑创生，演变的全过程。我们比对了所有名单，选择出其中出现频次最高的100个建筑。

考虑到图片排布的一致性，也为了使本书便于携带，我们将图书尺寸和格式设计成目前的样式。每个建筑都至少包含一张水平投影图（场地平面或建筑平面）、一张垂直投影图（立面图或剖面图）和一张立体图（轴测图或体量图）。由于涉及的建筑范围极广，包含了从单户住宅到大型商业项目等众多种类，因此书中并没有采用相同的制图比例来处理图形。恰恰相反，为了尽可能利用页面空间，我们缩放了图片大小并提供了比例尺作为参考，还为每个建筑配置了实景照片。为了减少有可能出现的失误，我们竭尽所能收集了手头所有资料，尽可能使用建筑师本人提供的或是建筑师作品集中出现的图片，实在没有权威资料的则至少选择了三个资料来源以绘制和验证图片。由于图书尺寸的原因，我们略去了地形要素，重点表现建筑概念、结构、空间和功能组织部分的内容。

毫无疑问，本书所选的100例建筑作品绝不是一份"终极名单"，也没法做到绝对科学。大部分建筑师都提供了超过100例的建筑方案，其中许多人还表示很难削减自己的名单。出于这个原因，同时也为了今后的进一步研究，我们将每位建筑师提供的名单作为附录编入书中。每一份清单的排列顺序都体现了建筑师本人的考量，同时也有许多人表示名单的排列顺序并不代表这些建筑在其心目中的排名。扎哈·哈迪德（Zaha Hadid）交付的名单是一个例外。非常不幸，她已离开了我们，无法再削减自己的清单了，所以我们将她提供的原始名单完整地呈现给大家。本书末尾的参考书目记录了研究过程中我们收集到的宝贵资源，可以协助我们对建筑项目作出进一步的调查研究。

最后需要指出，有不少令人尊敬的同事通过灵活或非正式的方式（如谈话）为本书的编制建言献策，还有很多人以类似方式对本书做出了贡献。我们所邀请的许多建筑师都做出了书面的回复，其他人则通过口述向我们提交了意见。也有不少建筑师无法回应，还有许多人无法取得联系。最后，希望本书可以被视为我们这一阶段工作的快照，展示出多样性的结果。

——Eui-Sung Yi
The Now Institute 主任

建筑 100

01 萨伏伊别墅
02 朗香教堂
03 巴塞罗那博览会德国馆
04 蓬皮杜中心
05 约翰逊制蜡公司总部
06 范斯沃斯住宅
07 萨尔克生物研究所
08 玻璃之家
09 拉图雷特修道院
10 环球航空公司候机楼
11 柏林爱乐音乐厅
12 毕尔巴鄂古根海姆博物馆
13 流水别墅
14 林地墓园
15 西格拉姆大厦
16 所罗门古根海姆博物馆
17 埃姆斯自宅(8号案例住宅)
18 莱斯特大学工程楼
19 法西斯宫
20 马赛公寓
21 栖息地67号
22 金贝尔美术馆
23 施罗德住宅
24 中银舱体大楼
25 邮政储蓄银行
26 马拉帕特别墅
27 悉尼歌剧院
28 鲁道夫馆
29 菲利普斯埃克塞特中学图书馆
30 AEG 透平机车间
31 罗比住宅
32 申德勒自宅
33 巴西议会大厦
34 包豪斯德绍校舍
35 香港汇丰银行大厦
36 格拉斯哥艺术学校
37 玛丽亚别墅
38 昌迪加尔议会大厦
39 斯德哥尔摩公共图书馆
40 代代木国立综合体育馆
41 仙台媒体中心
42 爱因斯坦天文台
43 利华大厦
44 钻石农场中学
45 玻璃屋
46 福特基金会总部大楼
47 慕尼黑奥林匹克体育场
48 盖里自宅
49 伊瓜拉达墓园
50 克朗楼
51 母亲住宅
52 洛弗尔健康之家
53 图根哈特住宅
54 贝克公寓
55 巴拉甘自宅与工作室
56 拜内克古籍善本图书馆
57 圣保罗艺术博物馆
58 布里翁墓园
59 孟加拉国国会大厦
60 斯图加特美术馆
61 西班牙国家罗马艺术博物馆
62 波尔多住宅
63 米拉公寓
64 赛于奈察洛市政厅
65 伦敦劳埃德大厦
66 阿拉伯世界研究中心
67 拉维莱特公园
68 柏林犹太人博物馆
69 代官山集合住宅
70 史密斯住宅
71 德国国家美术馆新馆
72 古堡博物馆
73 莱萨浴场
74 米歇尔广场公寓(路斯公寓)
75 鲁萨科夫工人俱乐部
76 棉纺织协会总部
77 斯塔尔住宅(第22号案例住宅)
78 布罗伊尔楼(惠特尼美国艺术博物馆)
79 屋顶加建
80 拉金公司行政大楼
81 菲亚特工厂
82 穆勒别墅
83 范内勒工厂
84 动态高效住宅
85 奎里尼·斯坦帕利亚基金会更新
86 六号住宅
87 贝尔拉赫证券交易所
88 甘布尔住宅
89 洛弗尔海滨住宅
90 都林展览馆
91 阿姆斯特丹市孤儿院
92 卡朋特视觉艺术中心
93 柏林自由大学
94 鲍斯韦教堂
95 虎塔婭艺术中心
96 柯布西耶中心(海蒂·韦伯住宅)
97 加拉拉特西公寓
98 中央贝纳保险公司大楼
99 维特拉消防站
100 横滨国际港客运中心

12 /

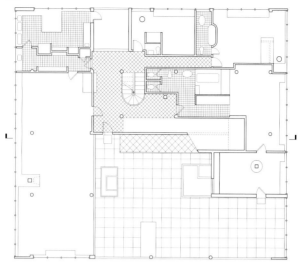

平面图

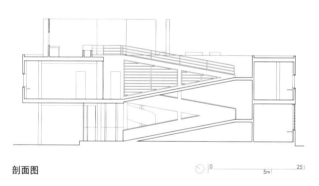

剖面图

1928-1931

北纬 48°55′28″

VILLA SAVOYE
LE CORBUSIER and PIERRE JEANNERET
Poissy, France

01

萨伏伊别墅

勒·柯布西耶与皮埃尔·让纳雷

法国，普瓦西

1928—1931 年；1963—1997 年修复

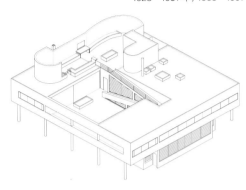

　　萨伏伊别墅清晰地表现出勒·柯布西耶的新建筑五要点（five points for a new architecture）：底层架空柱、自由的平面、自由的立面、水平带型窗以及屋顶上的花园。这栋建筑被认为是纯粹主义的宣言式作品，浴室、壁橱、楼梯和墙壁雕塑般的轮廓被附以鲜明的色彩，它们与建筑"白色画布般"简洁的外部几何形整体结构形成了清晰的对比。在这座乡村别墅的设计中，建筑师以玻璃和混凝土重新演绎了古典别墅的尺度与比例，诠释了"居住的机器"这一概念。建筑底层以汽车的转弯半径为尺塑为弧形；在其内部，一条连续坡道贯通各层直抵屋顶，营造出从地面到天空的连续建筑体验，这一点正是勒·柯布西耶"散步建筑"的极好示例。

1963-1997

东经 2°01′42″

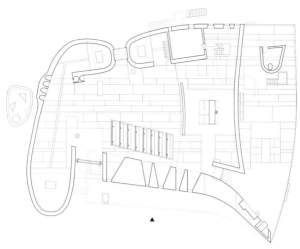

首层平面图

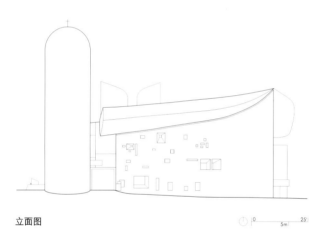

立面图

北纬 47°42′16″

CHAPELLE NOTRE-DAME DU HAUT
LE CORBUSIER
Ronchamp, France

02

朗香教堂
勒·柯布西耶
法国，龙尚
1950—1955 年

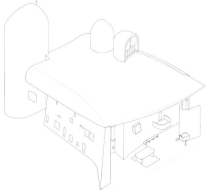

　　这座位于龙尚的朝圣教堂在建成后曾被谴责"像印象派般轻率"，但它却是整个 20 世纪最具代表性的宗教建筑之一。建筑的厚重体量呈现出飘浮般的错觉，但它的真实结构却轻盈得令人惊叹。向上收拢的巨墙如中世纪建筑般凝重，实际上是一座巨大的钢筋混凝土框架。建筑师在这面凹墙的水泥中掺入了之前小教堂（在第二次世界大战中被摧毁）的碎石，巨大的混凝土屋顶则"悬停"在墙体的上方。以巨墙为起始顺时针围绕建筑移动，便可看到两个外鼓的立面，三座高耸的采光塔矗立其间。建筑的第四个立面向内部凹进，为室外祭坛提供了空间上的围护，同时也起到了反射声音的作用。内凹的墙体中央是一个方形的"窗户盒子"，里面安放着历史悠久的木质圣母雕像。在内部，建筑地面顺着地形自然倾斜，向下漫步，便可抵达室外的祭坛广场。

东经 6°37′15″

1950-1955

16 /

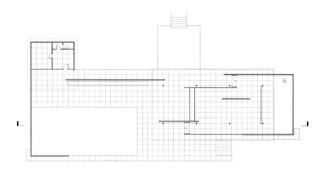

首层平面图

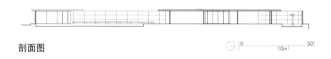

剖面图

GERMAN PAVILION / BARCELONA PAVILION
MIES VAN DER ROHE
Barcelona, Spain

03

巴塞罗那博览会德国馆

密斯·凡·德·罗

西班牙，巴塞罗那

1928—1929 年建造，1930 年拆毁，1986 年复建

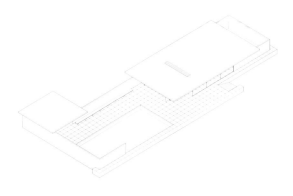

　　这座为 1929 年巴塞罗那国际博览会修建的临时建筑因其卓越的形式表现手法，被誉为最伟大的现代主义建筑杰作之一。展览馆的屋顶被八根镀铬的瘦长立柱支撑，表现出飘浮感。屋顶之下安置着一系列不规则布置的大理石墙壁，表现出玻璃般的光滑整齐，而落地玻璃窗则表现了镜面的反射效果；乳白色的磨砂玻璃墙则呈现出发光体一般的明亮感。整栋建筑看起来是理性功能主义的结果，却又在致敬古典建筑的正统。同时，这座建筑也展现出战后魏玛共和国的繁荣、具有理性的远见以及充满活力的现代性。原建筑在博览会结束后拆除，目前的德国馆是 1986 年重建的。

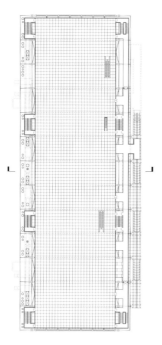

平面图

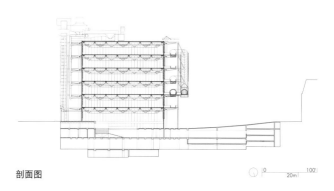

剖面图

北纬 48°51'39"

CENTRE POMPIDOU
RENZO PIANO and RICHARD ROGERS
Paris, France

04

蓬皮杜中心

伦佐·皮亚诺与理查德·罗杰斯

法国，巴黎

1971—1977 年

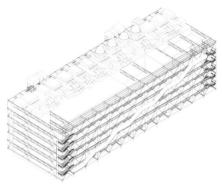

　　蓬皮杜中心本质上是一座由内而外的建筑，它颠覆了建筑各构件典型的组织关系。其外露的、色彩鲜明的循环和机械系统与外部结构内包含着巨大而多变的空间。在独特的建筑形态之下，设计师将内部的功能提升到艺术的层次，外部暴露的结构与机械构件使得建筑功能更加多变难测。各个功能在这栋建筑中相互交织：艺术博物馆、图书馆、文化中心、咖啡馆、商店和行政办公室各安其位。游客可以在封闭玻璃管道内的自动扶梯上远眺巴黎的城市轮廓（最初的设想是，建筑的层高可以根据策展的需求而抬高或降低）。可以说，蓬皮杜中心是一个公共广场——正如锡耶纳的坎波广场（Piazza del Campo）将市民们汇聚在一起——只不过这栋建筑的占地面积只有坎波广场的一半。

东经 2°21'08"

1971-1977

北纬42°43'46"

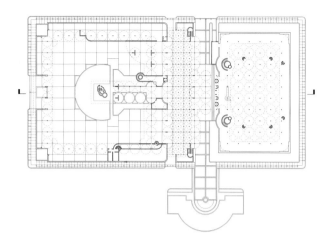

首层平面图

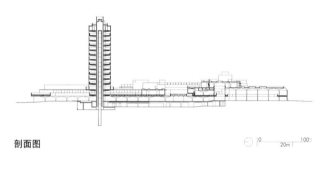

剖面图

西经87°47'29"

S.C. JOHNSON & SON HEADQUARTERS
FRANK LLOYD WRIGHT
Racine, Wisconsin, USA

05

约翰逊制蜡公司总部
弗兰克·劳埃德·赖特
美国，威斯康星州，拉辛
1936—1951 年

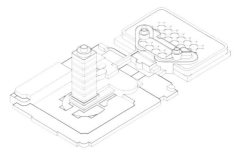

　　S·C·庄臣父子公司[1]总部蜿蜒的立面由一系列舒展的横向砖墙和派莱克斯玻璃管[2]构成，形式统一的外部界面围合着由钢筋混凝土树状预制立柱支撑的巨大垂直空间——"大工作间"。预制立柱的底部不足一英尺宽，却能支持其上部直径18英尺的圆形顶盖，顶盖之间的空隙则覆盖着玻璃管组成的采光顶。这些空心玻璃管使建筑顶部射入的光线发生漫反射，有效提升了工作空间的光线环境。方形的研究塔楼通过一条长廊与主楼连通，其中央是钢筋混凝土核心筒，各层楼板从核心筒悬挑而出，两层之间还设置了圆形的夹层空间。随着高度的抬升，各层楼板的面积也逐层轻微增大。虽然塔楼的原始结构看起来稳固而厚重，但是立面处理中通过增加"玻璃管窗"并削减砖材的面积，使这栋塔楼最终产生了半透明的视觉感受。

22 /

北纬 41°38'06"

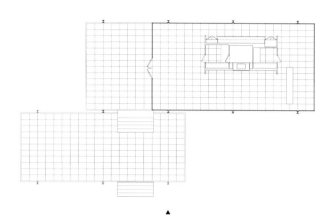

平面图

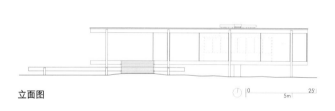

立面图

西经 88°32'09"

FARNSWORTH HOUSE
MIES VAN DER ROHE
Plano, Illinois, USA

06

范斯沃斯住宅
密斯·凡·德·罗
美国，伊利诺伊州，普莱诺
1945—1951 年

  这座周末别墅以单层的几何形体展现出最简单的居住理念，也印证了密斯的名言——"少就是多"。范斯沃斯住宅的概念构思非常简单——一个置于自然环境之中的玻璃住宅。它利用周围环境作为建筑外部的遮掩，而内部空间则被非凡的外表皮所包围。三个简单水平面（悬台、地面与屋顶）被外露的钢柱提升起来，结成了一种极开放、极精彩的构成关系；晶莹透亮的落地玻璃窗水平而舒展，衬托着外露的结构部件。一个包含浴室、厨房、多用空间、壁炉和储藏室的功能盒子，与小卧室中的储藏单元一起分隔了建筑的内部空间，它们是房屋内部仅有的两部分实体。尽管范斯沃斯住宅表现了现代主义的视觉特征，却因其对基座、立柱、楣梁等传统要素的重新演绎而被看作诠释古典建筑理想的杰作。

1945-1951

24 / 北纬 32°53′15″

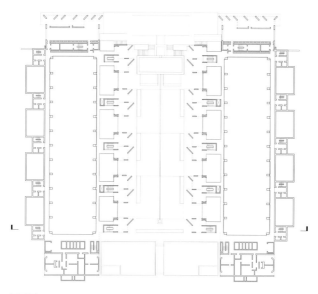

平面图

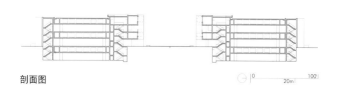

剖面图

SALK INSTITUTE
LOUIS KAHN
La Jolla, California, USA

07

萨尔克生物研究所
路易斯·康
美国，加利福尼亚州，拉霍亚
1959—1965 年

　　萨尔克生物研究所犹如一部静谧而巨大的研究机器，带着对大海的敬意将轮廓融入了海岸线中。康在这座建筑中将其"服侍空间"（servant）与"被服侍空间"（served）的组织策略与深邃的诗意捏合在一起。建筑用柚木及加入了火山灰的混凝土作为材料，以中央的大理石庭院为轴，[1] 呈现对称式的整体布局。源起于建筑入口的细水槽引导着人的目光穿过整个中央庭院，将视线导向远方无垠的大海和天际线。两侧楼阁的混凝土表皮赋予了庭院空间原初的朴素感，外突三角形空间的木质立面则朝着大海远眺。这座建筑有相当一部分功能被放在地下，自然光通过采光井照亮了地下层的实验室。

1959-1965

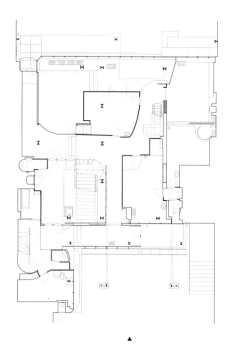

首层平面图

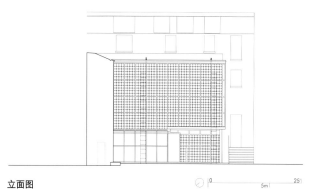

立面图

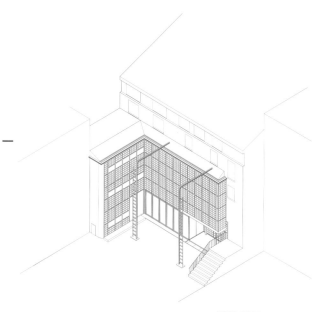

1928-1932

LA MAISON DE VERRE
PIERRE CHAREAU and BERNARD BIJVOET
Paris, France

08

玻璃之家
皮埃尔·沙罗与贝尔纳·贝福特
法国，巴黎
1928—1932 年

　　皮埃尔·沙罗与贝尔纳·贝福特合作设计的玻璃之家是现代建筑早期工业化分支中的先驱式作品。在这栋建筑中，设计者始终坚持使用钢材、玻璃与玻璃砖等工业材料与工业化固定方式；其内部装饰派艺术风格[1]的家居与装修同样出自沙罗之手。夜晚，建筑点亮灯光，多层玻璃砖立面就焕发出朦胧的光彩。这栋住宅被建在一处庭院之内，其3层高的建筑体量完全被一栋原有的四层公寓楼所包围，却依靠横纵交叉的宽法兰柱[2]与钢结构创造出自己的"自由平面"。内部的机械装置与建筑形成了一种独特的关系，通过折叠、滑动和旋转房间里的隔板与玻璃柜就可以改变室内的空间：例如，在白天营业时，一面能够旋转的穿孔板将让·达尔萨斯医生（Dr. Jean Dalsace）的私人楼梯掩藏起来，一层诊疗室内的病人却丝毫看不出这个精巧绝伦的装置。

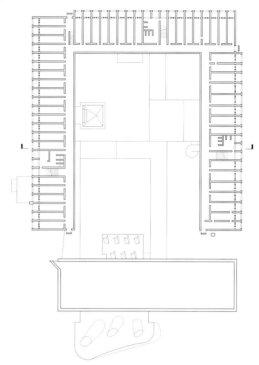

平面图

剖面图

CONVENT OF SAINTE-MARIE DE LA TOURETTE
LE CORBUSIER
Eveux-sur-l'Arbresle, France

09

拉图雷特修道院
勒·柯布西耶
法国，拉尔布莱勒的埃沃
1953—1960 年，1981 年复建

　　拉图雷特修道院位于里昂（Lyon）附近，是一座多明我会[1]僧院。在这栋建筑中，柯布西耶将关于仪式、地景、光线、声音和运动的各种设计构思整合在一起。跳跃的剖面与规矩的平面对抗着，二者的辩证关系为活跃的功能布置提供了依据：从平面来看这几乎是一座典型的修道院，而倾斜的地形和建筑的平屋顶却推动着柯布西耶在剖面设计中做出一系列前所未有的创新。沿着教堂底层被强调的毛坯表面，人被引导至位于斜坡顶部的建筑入口，它处于建筑体量的质心位置；修士的宿舍和独立的小礼拜堂布置为 U 形格局放置在入口上方，由此下转，就到达了收纳、用餐和祷告的空间；修道院窗户的韵律与式样则应和着设计合作者伊安尼斯·克塞纳基斯（Iannis Xenakis）[2]的乐谱结构。整栋建筑从里到外都以柯布西耶的"模度尺"为参照，协调统一了各个建筑元素。

1953-1960　　1981

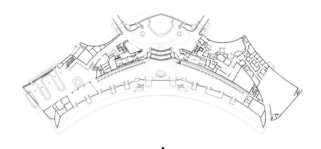

首层平面图

立面图

TWA FLIGHT CENTER
EERO SAARINEN
Queens, New York, USA

10

环球航空公司候机楼
埃罗·沙里宁
美国，纽约州，皇后区
1956—1962 年

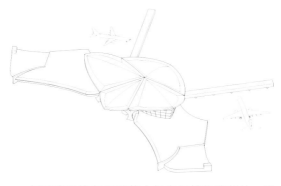

  与西班牙毕尔巴鄂的古根海姆博物馆相比，埃罗·沙里宁设计的环球航空公司候机楼采用了完全不同的塑形方式。[1]这座建筑如同一只张开羽翼的大鸟，好像随时会展翅腾飞。无论从视觉上或经验上来说，航站楼连续而蜿蜒的楼梯、平台和通道都代表着一种"飞行的精神"。主楼屋盖采用加强筋与混凝土现浇的结构方式，表现出复杂而立体的曲线；自主楼穿过轻微拱起的管状混凝土廊桥就到达了两翼延伸出的配楼，通过建筑师精心的设计，旅客步行到登机口的体验被缩短并弱化，同时加强了旅客到达后"宾至如归"的建筑感受。候机楼提供了一整套建筑环境设计，每处细节都重申并强化了沙里宁设计概念中的审美意图。

1956-1962

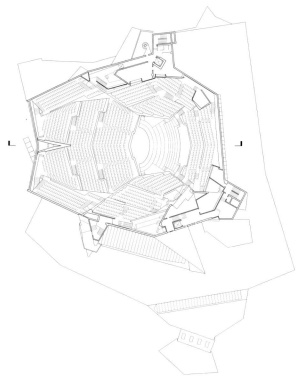

平面图

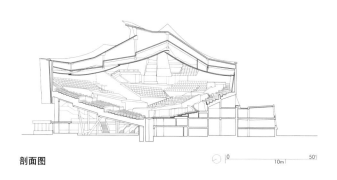

剖面图

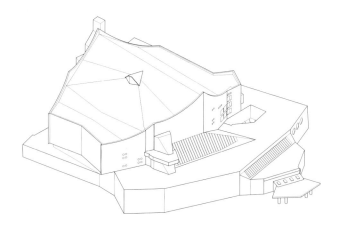

北纬 52°30'36"

BERLINER PHILHARMONIE
HANS SCHAROUN
Berlin, Germany

11

柏林爱乐音乐厅
汉斯·夏隆
德国,柏林
1956—1963 年

在设计柏林爱乐音乐厅时,夏隆声称要将音乐放在建筑的中心——无论从字面意思还是建筑形式而言他都做到了这一点——在这座音乐厅中,他第一次实践了自己发明的"葡萄园式"座位布局。乐团的席位布置在主厅的中心位置,观众席则层层叠叠地分布在乐池周围。无论坐在哪里,观众都觉得自己与演奏区非常接近,已成为这个音乐小团体的一部分。大厅的空间形状可以优化乐队演奏的声学效果,帐篷式的顶棚可以均匀地聚焦厅内的声音。虽然音乐厅的外观非常独特,却忠实地表现出内部的空间体量。大厅中央的"山峰"被玻璃屋顶笼罩,辅助空间则采用了混凝土屋盖——这些手法皆清晰地反映出夏隆"有机建筑"的设计观念。

东经 13°22'11"

1956-1963

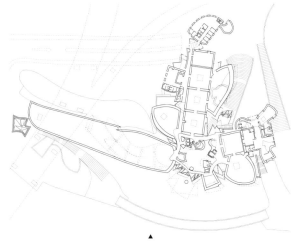

平面图

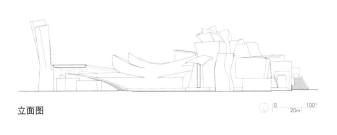

立面图

西经 2°56′00″

北纬 43°16′07″

GUGGENHEIM MUSEUM BILBAO
FRANK GEHRY (GEHRY PARTNERS, LLP)
Bilbao, Spain

12

毕尔巴鄂古根海姆博物馆
弗兰克·盖里（盖里建筑事务所）

西班牙，毕尔巴鄂
1991—1997 年

  毕尔巴鄂的古根海姆博物馆本身就是一座艺术品，它以众多自由曲面组合而成的复杂造型著称。自这座建筑之后，都市博物馆便开始在城市文脉中扮演一种全新的角色。这座建筑是毕尔巴鄂重建计划的一个组成部分，旨在更新这座老旧的工业城市并使其更加现代化。它嵌在旧城区的边缘，奔涌的形体横卧在河岸之上，一部分体量从大桥之下滑过，连接着远处的塔楼。高耸的塔楼作为标志物将上下两个城区联系在一起。正是由于 3D 软件建模技术（如 CATIA）的不断进步，设计师才可以表现出如此非凡的复杂曲线造型。这项技术开启了一个全新的时代，从此建筑师能够将设计意图直观地提交给建造商。一度萧条的工业城市毕尔巴鄂，因为古根海姆博物馆的成功得以恢复元气——至此，建筑作为一种极富价值的基础设施带动城市经济与社会转型的概念被越来越多的人所接受，这就是广为流传的"毕尔巴鄂效应"（Bilbao effect）。

1991-1997

36 /　　　　　　　　　北纬 39°54′23″

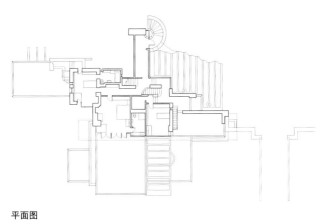

平面图

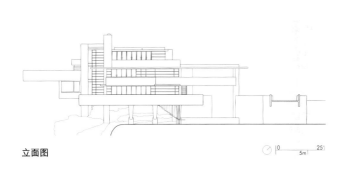

立面图　　　　　　　　　　0　　25′
　　　　　　　　　　　　　　5m

西经 79°28′05″

1934-1937

EDGAR J. KAUFMANN HOUSE / FALLINGWATER
FRANK LLOYD WRIGHT
Mill Run, Pennsylvania, USA

13

流水别墅 [1]
弗兰克·劳埃德·赖特
美国，宾夕法尼亚州，米尔兰
1934—1937 年

这栋周末别墅是赖特有机建筑概念最具代表性的作品。在此，赖特写就了一篇关于悬臂与声响、垂直与水平的散文诗。它的基座牢牢扎根于溪岩，露台却高高外挑，反抗着大地的引力；时而可见自然的裸岩从地板和墙壁中迸发而出，时而又听见溪流的潺潺水声，只闻其声却不见其貌，这是建筑在同自然嬉戏；由一部阶梯向下行进，终于见到了使建筑得名的溪流。建筑内外的各元素——铁框的窗户、石支柱、混凝土的露台，都从不同的尺度表现出连续性。

38 /

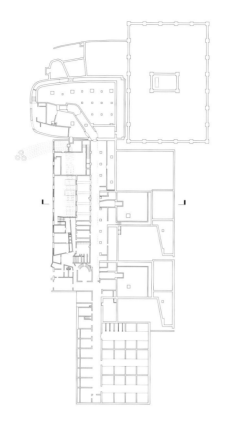

平面图

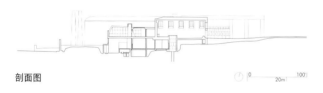

剖面图

1915-1940

THE WOODLAND CEMETERY
ERIK GUNNAR ASPLUND and SIGURD LEWERENTZ
Stockholm, Sweden

14

林地墓园
埃里克·贡纳尔·阿斯普隆德与西古德·莱韦伦茨

瑞典，斯德哥尔摩

1915—1940 年

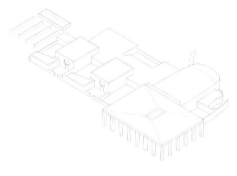

　　林地墓园也被称作森林公墓（Skogskyrkogården），由埃里克·贡纳尔·阿斯普隆德与西古德·莱韦伦茨合作设计，二人在 1915 年通过竞标获得了这个项目。这项杰作完全遵循斯堪的纳维亚"教堂花园"的安葬传统，在森林的自然美景中开辟出一系列路径与景观。树林、小礼拜堂、火化场、喷泉和辅助建筑散布各处，叙述着对于宁静、希望与安憩的渴望。这项作品是阿斯普隆德自早期北欧古典主义（Nordic classicism）转向现代主义的变法之作（莱韦伦茨在完成了复活教堂建筑群后便离开了该项目）。建筑师无所不包地设计了所有的元素——从道路铺装到固定卡具，营造出一种迷人且富有感染力的完整自然体验。林地墓园在 1994 年被联合国教科文组织列入世界遗产名录。

40 /

北纬 40°45′30″

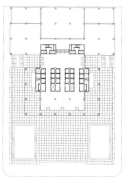

首层平面图

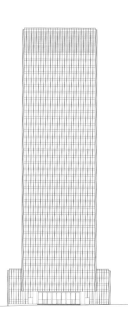

立面图

西经 73°58′20″

SEAGRAM BUILDING
MIES VAN DER ROHE and PHILIP JOHNSON
New York, New York, USA

15

西格拉姆大厦

密斯·凡·德·罗与菲利普·约翰逊

美国，纽约州，纽约市

1954—1958 年

密斯设计的西格拉姆大厦蕴含着简单与优雅的力量，立面中比例恰当的金属格栅被提升到完美的境界。这座建筑是一家加拿大制酒公司位于美国的总部，也是建筑师在美国设计的第一个大型商业项目。修长的塔楼耸立在一片宁静的粉红色花岗石广场上，广场西侧的拐角处设置了两面宽阔的水池，基址东端的一片墩座上有两家风格独特的餐厅（目前那里已经发生了很大变化）。

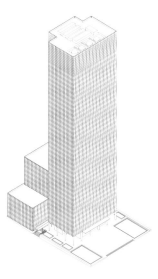

密斯的板楼面对着由麦克金、米德与怀特所设计的新古典主义的纽约网球私人会所，两栋建筑在对称性、比例、材料、细节上皆发生了引人入胜的奇妙对话。大厦的建筑表皮表现出一种强有力的构造逻辑，讲述着建筑的永恒与庄严，垂直的宽翼工字钢与横向线条织就出建筑的黑古铜色立面；包覆着混凝土的钢梁柱被立面围合，使得建筑真正的结构不得而见。罗曼式的前厅使用产自佛得角（Verde）的仿古大理石板，安静而清晰地讲述着细节与工艺对于这栋建筑的巨大意义。

1954-1958

北纬 40°46′59″

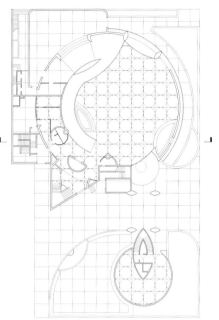

首层平面图

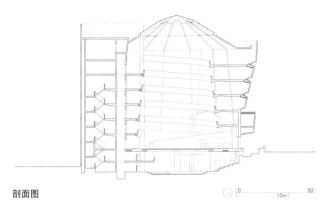

剖面图

西经 73°57′32″

SOLOMON R.GUGGENHEIM MUSEUM
FRANK LLOYD WRIGHT
New York, New York, USA

16

所罗门古根海姆博物馆
弗兰克·劳埃德·赖特
美国，纽约州，纽约市
1943—1959 年

所罗门古根海姆博物馆的设计方案极为大胆。姑且不论它如此自顾自地矗立在纽约稠密的都市环境中，这座建筑倒椎形的体量也与其所处的街区脉络形成了鲜明的对比；不仅如此，它看起来似乎还要闯过中央公园，与比它大得多的大都会艺术博物馆（Metropolitan Museum of Art）争夺不朽的地位。古根海姆博物馆统一了形式、结构和空间，可以称得上是赖特"有机建筑"概念的城市版本。纯白色的倒置截头锥形体[1]采用加筋混凝土结构，从形式相似的球形基座中浮现出来。建筑内部的螺旋式连续坡道——这条本来计划从顶部发起的长廊——从画廊的壁龛中悬挑而出，它不但塑造了建筑的倒锥形体，也决定了中庭的形式，中庭顶部则被圆形的玻璃穹顶所覆盖。倒锥体表面的切口实际上是结构的变形缝，光线透过其中的遮光板映射进画廊，明暗之间诉说着循环不休的螺旋画廊的本质。

44 /

北纬 34° 01' 49"

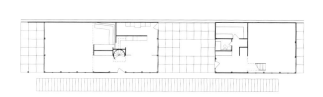

首层平面图

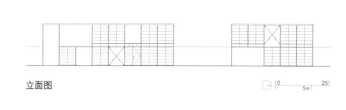

立面图　　　　　　　　　　　　　　　　　　　　0　　5m　　25'

西经 118° 31' 10"

1945-1949

17

THE EAMES HOUSE
CASE STUDY HOUSE NO.8
CHARLES and RAY EAMES
Pacific Palisades, California, USA

埃姆斯自宅（8号案例住宅）[1]

查尔斯·埃姆斯与雷·埃姆斯

美国，加利福尼亚州，宝马山花园

1945—1949年

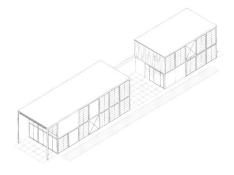

埃姆斯夫妇是一对典型的20世纪美国设计师，他们的业务范围极其广泛：从玩具、纺织品、工具、家具到书籍、电影和策展，各个方面都有所涉及。8号案例住宅是他们基于第二次世界大战（以下简称"二战"）期间研发的预制材料和建造技术设计出的一栋原型住宅，埃姆斯夫妇希望以此方式将他们的工业设计技能带入建筑领域。这栋房屋按照功能分为两部分——住宅与工作室——二者一同匍匐在陡峭山坡的轮廓之中。背靠斜坡的挡土墙构成了建筑的一面边界，而暴露在外的钢结构框架则提示出这栋建筑的平面和剖面是按照模块化组织的方式拼合的。整栋房屋被网格化的窗户和玻璃门围合起来，色彩明丽的不透明面板则画龙点睛一般穿插在立面之中。

46 /

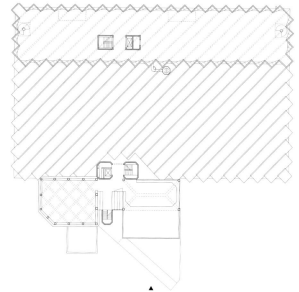

平面图

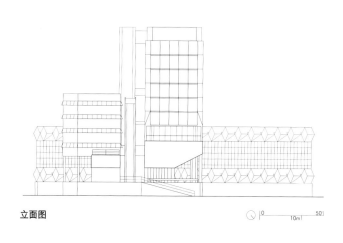

立面图

西经 1°07'25"

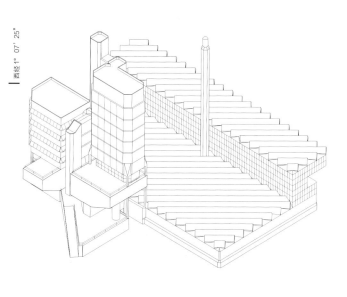

北纬 52°37′13″

18

LEICESTER UNIVERSITY ENGINEERING BUILDING
JAMES STIRLING and JAMES GOWAN
Leicester, England

莱斯特大学工程楼
詹姆斯·斯特林与詹姆斯·高恩
英国,莱斯特
1959—1963 年

在莱斯特大学工程楼的设计中,斯特林与高恩戏耍般地将小体量与大体量捏合在一起,建筑宏大却不失精巧,现代建筑样式中却又流露出哥特建筑的气质。斯特林的设计意图是将自己学生时代就试验过的"学院机器"以更全面的方式呈现出来。从表面上看,这栋建筑基于功能的安排塑造出各部分形式:开放式的工作空间设置了45°斜切的锯齿形天窗,无柱的实验室里设置了照明控制,办公大楼也具备相当不错的采光和视野,礼堂则坦率地表现了封闭的体量感。整栋建筑清晰地表现出构成主义(constructivist)与工业化的浪漫主义(industrial romanticism)特征。不规则的透亮棱柱形元素在建筑中反复出现,它们看起来甚至有些古怪,却起到了联系和区别建筑各部分的作用。

1959-1963

48 /

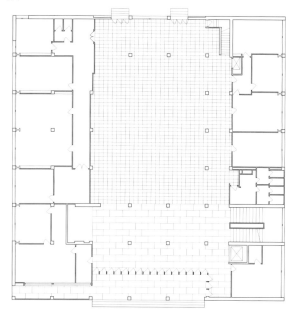

首层平面图

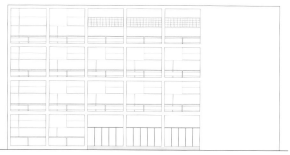

立面图

1932-1936

CASA DEL FASCIO
GIUSEPPE TERRAGNI
Como, Italy

19

法西斯宫
朱塞佩·泰拉尼
意大利,科莫
1932—1936 年

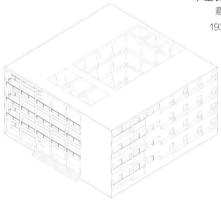

　　朱塞佩·泰拉尼参照罗马法尔尼斯府邸[1]的平面布局设计了法西斯宫。在意大利的法西斯政权全面转向更具影响力的新古典主义风格之前,这项作品代表着意大利现代主义建筑的高峰。构成建筑的一系列元素——框架、墙壁和玻璃以完全抽象的形式在建筑组织中不断地变化位置关系,各个元素的意义也因此而发生了一系列转变。根据建筑功能和防晒的需求,四个沿街立面在建筑表皮和空间深度上皆体现出不同的变化。覆盖着玻璃屋顶的"中央庭院"可以看作室外广场在建筑内的延伸,十六扇玻璃门直接开向中庭,这最初是为了让人们可以从建筑物内部发起一场城市游行。

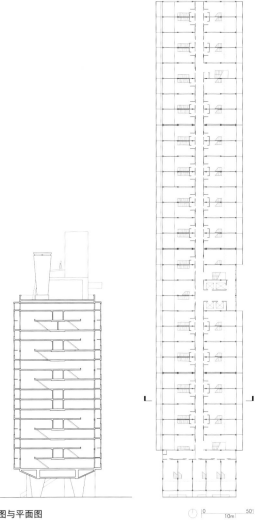

剖面图与平面图

北纬 43°15′42″

L'UNITÉ D'HABITATION
LE CORBUSIER
Marseille, France

20

马赛公寓[1]
勒·柯布西耶

法国,马赛
1945—1952 年

　　共有五座城市曾建设过"居住单位",这代表着勒·柯布西耶已将"建筑五要点"提升到了城市的尺度范围。最初的"居住单位"建造于马赛,这栋多功能集合住宅如同一艘远航的游轮,反映出柯布西耶基于战后法国的需求提出的社区生活愿景。这是一座典型的"公园中的塔楼",允许其居民在足以容纳 1600 人的建筑体量下面自由穿行;屋顶上则布置了数量众多的公共空间,简直就是一个独立的小城市。这栋集合住宅中包含多种居住单元——正如一张著名的照片所展现的概念:一只手正在将一个琴键般的体块插入一套方格框架。建筑中还采取了诸如交叉通风、双层通高空间等一系列策略,并在建筑立面中设置了很多的阳台。由于每 3 层楼只需要 1 条走廊,因此可以最大限度地提升自然通风和景观视野,空气循环系统也因此被极力缩减到最低水平。

1945-1952

东经 5°23′47″

北纬 45°29′60″

标准层平面图

剖面图

HABITAT '67
MOSHE SAFDIE
Montréal, Canada

21

栖息地 67 号
摩西·萨夫迪
加拿大，蒙特利尔
1967 年

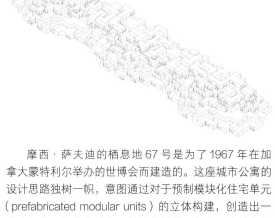

  摩西·萨夫迪的栖息地 67 号是为了 1967 年在加拿大蒙特利尔举办的世博会而建造的。这座城市公寓的设计思路独树一帜，意图通过对于预制模块化住宅单元（prefabricated modular units）的立体构建，创造出一种全新的三维住宅（由于结构荷载差异极大，建筑上部的模块单元与下部单元从形式和结构上都有很大区别）。为了在不牺牲郊区怡人风景环境的前提下尽可能提高建筑密度，设计师将住宅单元组织为三组连续的"金字塔"形体，并且单独设置了内部的交通筒。盒子般的住宅层峦交错，曲折变化中流露的间隙空间营造了一种有机的、乡村田园诗般的人居环境，生机盎然的城市花园遍布建筑周围。栖息地 67 号是一栋高效的集合住宅，也是一座现代的城市山居。

— 1967

54 /

北纬 32°44'56"

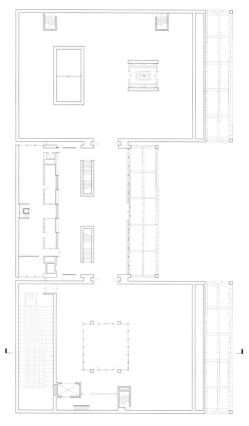

平面图

剖面图

西经 97°21'54"

KIMBELL ART MUSEUM
LOUIS KAHN
Fort Worth, Texas, USA

22

金贝尔美术馆
路易斯·康
美国，得克萨斯州，沃斯堡
1966—1972 年

　　金贝尔美术馆是康将其学院派[1]艺术功底与自己对原始体量、光影的当代探索融为一体的完美之作。主体由五跨平行的筒形拱组成（第六跨拱是门廊），建筑表面贴装了大理石板，精心栽种的树林安静地环绕在建筑周围。在这片风景如画的土地上，康以自己惯用的形式语汇创造了这栋美术馆——强烈的物质性、低调的轮廓、对称的布局，这是一座安静而亲和的古典主义纪念碑。然而一旦你步入其中，就会发现处处都体现着建筑师绝妙的构思：错置的庭院、出乎意料的功能排布、灵活的交通组织以及贯通建筑的采光缝，其空间与细节的复杂程度着实令人震惊。最具代表性的是建筑的天窗，它的构件清晰地标示出筒形拱顶的劈缝以及其上的纵脊。这个极具匠心的光线过滤装置已经成了金贝尔美术馆的标志性符号，以至于策展人全都使用建筑内的自然光线布展；同时，这栋建筑中石材和混凝土巧妙的组合方式也是当代建筑学习的经典范例。

1966-1972

平面图

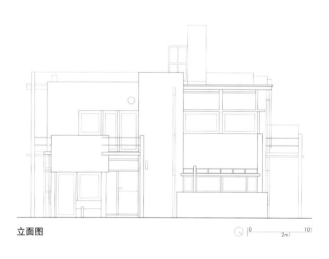

立面图

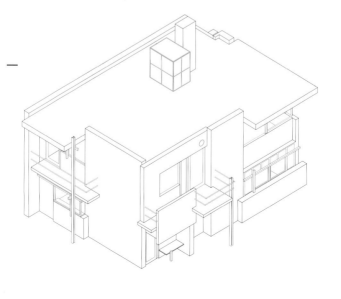

23

SCHRÖDERHUIS
RIETVELD SCHRÖDER HOUSE
GERRIT RIETVELD
Utrecht, The Netherlands

施罗德住宅
赫里特·里特韦尔
荷兰,乌德勒支
1924年,1984—1985年重建

施罗德住宅是荷兰风格派运动[1]的代表性作品,坐落于乌德勒支的一列早期联排住宅的末端。其割裂的立方形体展现出"去物质化"[2]的特征,一系列复杂的线条与平面组织颠覆了传统的住宅形态。经过重新演绎的建筑元素在室内和室外穿插延伸,模糊了内外的界限,导致内部与外部的形式几乎没有任何差别。为了强调建筑的直线关系,建筑师将所有窗户的开启方向都设置为垂直于所属立面,同时,整栋房屋在各个尺度都是由不同大小的彩色平面构成。建筑中的钢结构部件被附以厚重的深色,与浅色的墙壁产生了鲜明的对比关系。室内大多是动态而多变的开放空间,没有传统的房间分隔;建筑师还以可滑动的墙壁划分了各个功能区域,进一步增强了空间的流动性。

1984-1985

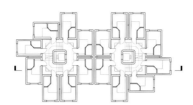

平面图

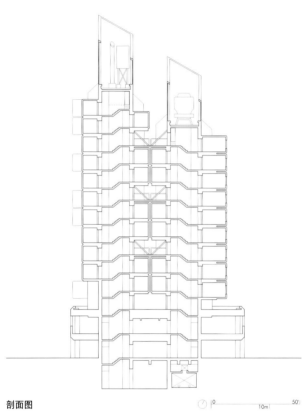

剖面图

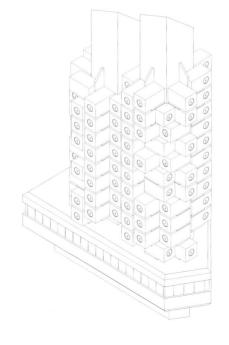

北纬 35°39′57″

NAKAGIN CAPSULE TOWER
KISHO KUROKAWA
Tokyo, Japan

24

中银舱体大楼
黑川纪章
日本,东京
1970—1972 年

中银舱体大楼是日本的新陈代谢派[1]硕果仅存的一项案例,是日本战后文化复兴与乐观主义精神的象征;同时,它还代表了一种在晚期现代主义中普遍蔓延的建筑狂想:固定的结构框架与可灵活装配的预制单元部件。这座多功能的办公住宅楼起初是为了接待出差在外的企业白领而设计的,共有14层高(下面两层是裙房),楼梯和电梯被布置在混凝土核心筒中,可拆卸的钢制模块化"胶囊"则被安装在核心筒外围。由于楼梯平台的位置灵活而多变,140个"胶囊"得以穿插错落地排布起来,建筑外形也因此充满动感。虽然预制"胶囊"单元的内部空间最初是按照容纳一间办公室或工作间的要求做出的设计,但也可以作为家庭的居住空间来使用。按照这栋建筑最初的设计构想,使用者可以随时用新预制单元更换陈旧的单元部件,但由于建筑已经过于老旧,再加上装配程序非常复杂,目前只能从建筑顶部自上而下地逐个拆除各个单元体块了。

1970-1972

东经 139°45′48″

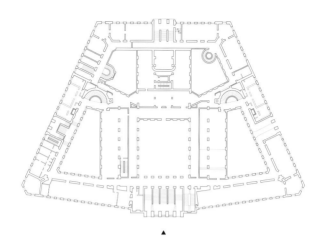

首层平面图

立面图

1903-1912

| 北纬 48°12′36″

POST OFFICE SAVINGS BANK
OTTO WAGNER
Vienna, Austria

25

邮政储蓄银行
奥托·瓦格纳

奥地利，维也纳
1903—1912 年

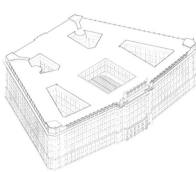

　　瓦格纳设计的邮政储蓄银行是一座超然独步的建筑，它站在 19 世纪新古典主义（neoclassicism）的身后，挺立在新艺术运动（Art Nouveau）的风口浪尖，昭示了即将来临的 20 世纪现代主义建筑。外部的铆钉大理石板、全玻璃的营业大厅、外露的铸铁支柱、晶莹剔透的中庭（最初是由桅杆和缆索悬吊起来的）、办公室内部的可移动隔板、铝制空气调节系统（通过瓦格纳的处理，机械部件甚至都表现出了纪念性），邮政储蓄银行所展现的设计手法都大大领先于其所处的时代。瓦格纳摒弃了众多维也纳建筑中虚假浮华的表皮装饰——作为一栋大型的公共建筑，邮政储蓄银行除了顶部的花环与山花雕像座以外（由奥斯马·施因科维茨设计）几乎没有任何装点，而这些仅有的装饰物也出人意料地采用了铸铝材料。建筑内的大多数照明装置和家具设计也出自瓦格纳本人之手，焕发出与建筑相得益彰的优雅美感。

东经 16°22′52″ |

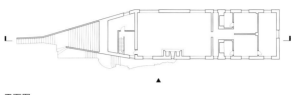

平面图

剖面图

立面图

— 1938-1940

CASA MALAPARTE
ADALBERTO LIBERA
Capri, Italy

26

马拉帕特别墅
阿达尔贝托·利贝拉

意大利，卡普里

1938—1940 年

  马拉帕特别墅是一栋清晰而独特的建筑，它匍匐在卡普里崎岖而多石的悬崖上，俯瞰着第勒尼安海。你可以将它理解成一部楼梯，也可把它看成一块顽石；它既是一片令人眩晕的平台，也是一处避难所。根据阿莱西娅·罗西塔尼·祖克特（Alessia Rositani Suckert）的说法，"这栋建筑虽说出自阿达尔贝托·利贝拉之手，但实际的设计者是其主人古奇奥·马拉帕特（Curzio Malaparte）。他曾按照自己的想法设计了一座极富个性的、表现力很强的住宅，并为它起名为'Casa Come Me'，字面意思就是'像我一样的房子'，最终马拉帕特将设计草图交到了绘图员翁贝托·博内蒂（Umberto Bonetti）手中"。人们可以坐船登岛，从高耸的悬崖边登上数不清的台阶，就来到了巨大楔形楼梯的基座——这本身便是主人流亡利帕里岛(lipari)的自传性隐喻——登上楼梯后你就到达了宽阔的平台，穿过风帆般飘扬的矮墙，一望无垠的大海尽收眼底。在这戏剧性的场景之下便是住宅的本体，每个独立房间的窗户都经过了精心的安排，是那辽远海面的取景框。

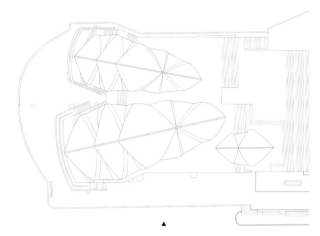

屋顶平面图

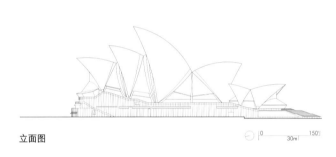

立面图

SYDNEY OPERA HOUSE
JØRN UTZON
Sydney, Australia

27

悉尼歌剧院
约翰·伍重
澳大利亚, 悉尼
1957—1973 年

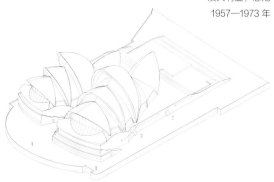

　　1957 年，伍重力拔头筹赢得了悉尼歌剧院的方案竞标——毫无疑问，它是整个 20 世纪最具标志性的建筑之一。歌剧院位于悉尼港一片突出的半岛上，优雅的阶梯式基座引导着观众进入一系列表演空间；观众也可在休息厅驻足，欣赏悉尼港的优美风光。基座之上的壳体结构看起来像是画笔的自由勾勒，实际上却是严谨的球体几何截面。依照伍重与奥韦·阿吕普（Ove Arup）合作开发的施工技术，首先立起预制混凝土尖拱，然后再将预制的"壳体"面板安装在上面；所有的预制件都出自一套通用模板。起初伍重希望通过剧院的声学特性来决定"外壳"的形状，但最终没能实现；[1] 不过，目前的建筑形态为布置剧场内的机械装置提供了充足的空间，宽阔的顶部还可以容纳一层小阁楼。

1957-1973

66 /

北纬 41°18'32"

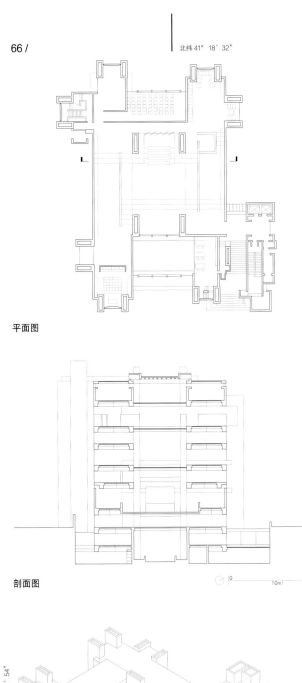

平面图

剖面图

西经 72°55'54"

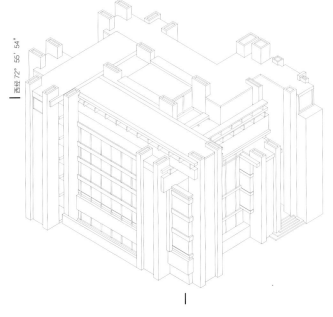

RUDOLPH HALL
Yale Art and Architecture Building
PAUL RUDOLPH
New Haven, Connecticut, USA

28

鲁道夫馆
保罗·鲁道夫
美国,康涅狄格州,纽黑文
1958—1964 年

耶鲁大学艺术与建筑系馆(现称为鲁道夫馆)是一座结构复杂而大胆的粗野主义(brutalist)建筑,其地上七层、地下两层的体量之内居然设置了37个不同标高的平面。这栋建筑无疑是保罗·鲁道夫对现代主义建筑及其根源最复杂、最精妙的申明性作品。建筑平面中显示出赖特的自由风格,而剖面设计则流露出皮拉内西[1]的风骨(这是保罗·鲁道夫的标志性设计手法)。建筑的外观着重体现出遍插着横梁的混凝土柱墩的垂直性,加厚的水平板则在梁与柱的空隙中交叉穿梭;鲁道夫馆通常被认为是在拥挤的拐角地段设计建筑时必须参考的典型范例。

1958-1964

68 /

北纬 42°58′45″

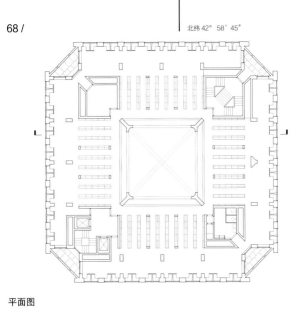

平面图

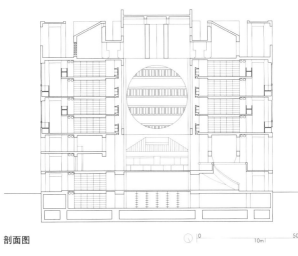

剖面图

西经 70°56′58″

PHILLIPS EXETER ACADEMY LIBRARY
LOUIS KAHN
Exeter, New Hampshire, USA

29

菲利普斯埃克塞特中学图书馆
路易斯·康
美国，新罕布什尔州，埃克赛特
1965—1971 年

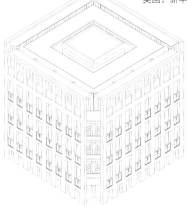

埃克塞特中学图书馆集中体现了康对于历史纪念性与现代主义的对位、个体与集合的辩证、光与影的表达等一系列概念的思考。阳光通过顶部侧窗的折射照亮了高达四层的中庭空间，各层的陈列书架围绕着中庭铺展蔓延，透过中庭内部四面巨大的混凝土圆洞可以看到学生们在各层书库中游走。外立面的砖墙被玻璃窗和柚木板分隔，形成了比例协调的网格；建筑四周专门设置了小阅览间，为了维护私密性及提高学习专注度，在与视线齐高的位置都安装了木质百叶窗。这座"图书大教堂"体现出康对于"静谧与光明"（silence and light）这一主题的偏爱，在其后设计的金贝尔美术馆和耶鲁大学美术馆中，康以不同的表达方式延续了这个设计主题。

1965-1971

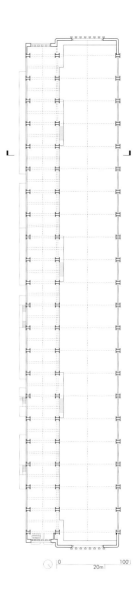

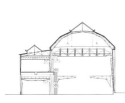

剖面与首层平面图

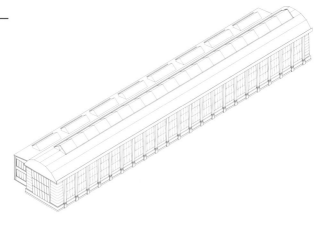

1908-1909

AEG TURBINE FACTORY
PETER BEHRENS
Berlin, Germany

30

AEG 透平机车间
彼得·贝伦斯
德国，柏林
1908—1909 年

贝伦斯的透平机[1]车间是德国现代建筑的先驱。在这栋建筑中，贝伦斯与结构工程师卡尔·伯纳德（Karl Bernhard）合作，将各种工业化材料组织成一栋建筑。内倾的现浇混凝土山墙、富于序列感的立面分隔、非对称的三铰钢框架都包含在二层高的简洁形体之内，若不是外露的钢结构间设置了能够采光的内倾玻璃幕墙，这座车间乍一看就是一座仓库。混凝土山墙自下而上逐渐向内倾斜，与垂直的钢柱形成鲜明的对比（此趋势在建筑角部尤为明显）。墙壁的视觉厚度因此被极大地削弱，实心的混凝土承重墙看起来似乎只起到围合的作用，一点也不像是结构体。钢结构固定在建筑基座上面，裸露的巨大铰链支撑起连续的钢制上横梁（entablature）。透平机车间的内部空间比照了巴西利卡的样式，将两层高的平行侧廊安置在高耸而细长的中庭两侧，古典的纪念性与机械的现代主义在建筑中交汇后形成了独特的面貌——这座建于 20 世纪初的杰作的确是一栋名副其实的"工业化大教堂"。

北纬41°47'23"

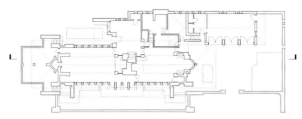

首层平面图

剖面图

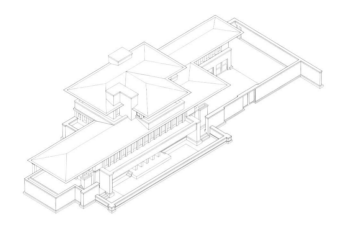

西经87°35'45"

1908-1910

FREDERICK C. ROBIE HOUSE
FRANK LLOYD WRIGHT
Chicago, Illinois, USA

31

罗比住宅

弗兰克·劳埃德·赖特

美国，伊利诺伊州，芝加哥

1908—1910 年

赖特为弗雷德里克·C·罗比一家设计的这栋住宅共有三层，还配有一间车库和一片停车场。该建筑不仅展现了"草原住宅"的所有特征，还浓墨重彩地表现了建筑中的水平要素（正是这一点让罗比住宅声名远播）。为了强调出水平性，所有墙体都以薄罗马砖为材料，同时还采用了两个独特的施工细节：砖块之间的竖缝使用了深红色砂浆，而水平方向则以醒目的暖白色砂浆勾缝，各层砖块间的抹灰清晰利落，在墙上勾画出一条条清晰可见的阴影。屋内的公共空间（起居室和餐厅）被安置在朴实的基座之上，向外望去便可见安静的街景，定制的彩色玻璃窗则为房间营造出丰富的光影效果；起居室与餐厅的交接位置设置了一座巨大的砖构壁炉，两侧空间都能够享受到温暖的炉火，垂直的烟囱直贯建筑的屋顶；从台球室和儿童游戏室向屋外漫步，便进入了一处被墙壁围合的庭院，安静的庭院与平缓的围墙从水平和垂直两个方向将建筑与一旁的街道分隔开来。

74 /

北纬 34° 05' 11"

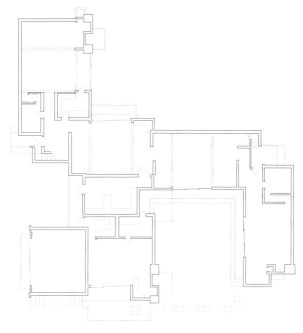

首层平面图

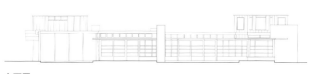

立面图

西经 118° 22' 20"

— 1921-1922

SCHINDLER HOUSE
RUDOLF SCHINDLER
West Hollywood, California, USA

32

申德勒自宅
鲁道夫·申德勒
美国，加利福尼亚州，西好莱坞
1921—1922 年

　　这栋申德勒设计的住宅是南加州现代主义建筑的开山之作，其室外空间相互映衬，十分出彩。建筑师并没有采用传统郊外别墅的"楼阁"形式，而是将建筑的空间与功能均匀分配在整片场地上面，让实体与空隙相交、封闭与开放共存、景观与建筑相映。这栋建筑的形状好似缺了一角的风车，整体布局由三处庭院和三翼房屋构成（住宅可供两户人同时使用），大部分室内空间仅由框架、滑动门和檐篷分隔，只在主人的私密空间周围设置了封闭的混凝土墙；每一翼房屋的入口都以壁龛的形式加以强调，院落中的火塘则为室外空间带来了温度与归属感。申德勒还在该建筑中实践了自己对公共居住模式的规划：受邀短住的客人拥有独立的卧室与花园，每户的入口旁边也布置了各自的休憩凉台，互不打扰，只有厨房是与主人合用的。

76 /

平面图

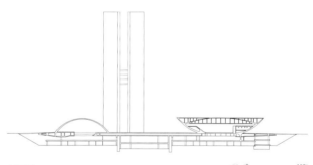

剖面图

西经 47°51′51″

南纬 15°47′58″

NATIONAL CONGRESS, BRAZIL
OSCAR NIEMEYER
Brasília, Brazil

33

巴西议会大厦
奥斯卡·尼迈耶
巴西,巴西利亚
1958—1960 年

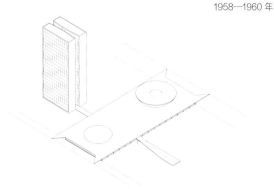

　　巴西议会大厦是奥斯卡·尼迈耶的杰作。它位于卢西奥·科斯塔[1]为巴西利亚新城（new city of Brasília）所作的飞翼状规划平面的轴线末端，是巴西法制与民主制度的象征。尼迈耶将科斯塔原始草图中的元素重新组织，把两个互相连接的矩形"石柱"（立法院办公楼）放置在桌面般的"基座"后面。一横一竖，巨大的建筑体量在波光粼粼的水面映衬之下极为协调；一仰一覆，两个碗形截面球体安静地停歇在水平基座上：较大且正置的是众议院会议厅，小而倒扣的是参议院会议厅；折返的长坡道在平面中形成了一条宏大的轴线，人们沿着坡道就可到达屋面。从远处望去，这栋建筑的几何形体如同静物画一般宁静而纯粹。

1958-1960

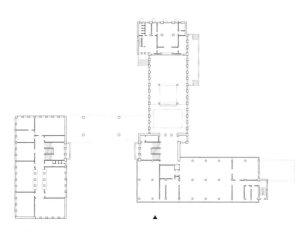

平面图

立面图

1925-1926

北纬 51°50′23″

BAUHAUS DESSAU
WALTER GROPIUS and ADOLF MEYER
Dessau, Germany

34

包豪斯德绍校舍

瓦尔特·格罗皮乌斯与阿道夫·迈耶

德国，德绍

1925—1926 年

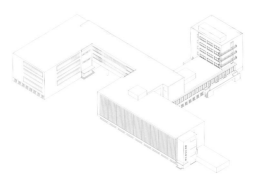

　　包豪斯校舍代表了一种全新的建筑设计思路，其形式的基础是对于美术、工艺和工业技术（包括其工具、材料和元素）的并行研究，这些成果最终以建筑的形式统合为一，并在建筑中达到了顶峰。"现代性"究竟意味着什么？这项宏大议题的答案就是包豪斯校舍。在当时的情况下，魏玛共和国并不信任包豪斯学校里那些颠覆性十足的先锋派——德绍校舍作为一种正式的"建筑风格"正好消除了这种疑虑。建筑的平面清晰地反映出格罗皮乌斯对于功能的规划：巨大的车间安装着工业化的玻璃幕墙，宽阔的公共楼梯将车间与底层的礼堂与食堂联系起来，底层架空的办公楼则把车间和教学楼连为一体。建筑中的每个分区都基于各自功能设置出独立的结构体系和窗户样式。

东经 12°13′38″

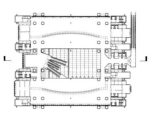

首层平面图

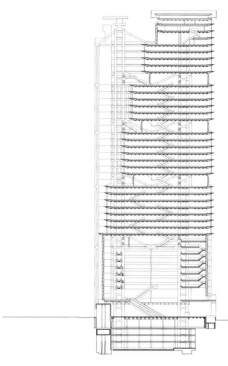

剖面图

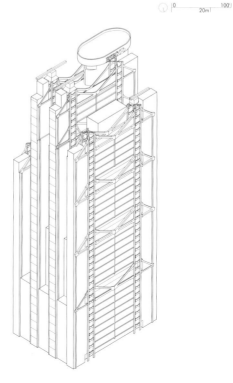

北纬 22°16′49″

35

HONGKONG AND SHANGHAI BANK HEADQUARTERS
NORMAN FOSTER (FOSTER & PARTNERS)
Hong Kong, China

香港汇丰银行大厦
诺曼·福斯特（福斯特建筑事务所）

中国，香港
1979—1986 年

汇丰银行大厦是诺曼·福斯特设计的第一栋高层塔楼，然而建筑师却立志将其设计成"世界上最好的银行"。以往的银行建筑大多是"庄严而封闭的堡垒"，而汇丰银行大厦却是一栋"透明的机器"；它不仅重新定义了银行建筑的形式，还前所未有地展现了"预制可插入式建筑"的广泛适应性（prefabricated plug-in architecture，这一概念兴起于 20 世纪 60 年代，却几乎从未被实现过）。四组管状的结构支架托举着五组两层楼高的钢桁架，桁架上面悬吊着五个楼层块体。三段塔楼依附着支撑结构逐步攀高，高度分别是二十九、三十六和四十层。为了解决采光问题，建筑师还发展出一套独特的光线折射装置，这些"折光斗"（sun scoop）位于塔楼中庭的侧方，突起的玻璃板将阳光向下折射，照亮了建筑底部的公共广场。汇丰银行大厦的内部空间极其丰富，不同规模的自动扶梯长廊和适应性极强的楼层设计向人们证明了预制建筑的可行性。

东经 114°09′34″

1979-1986

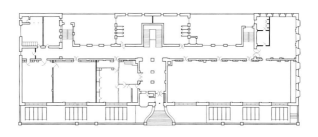

首层平面图

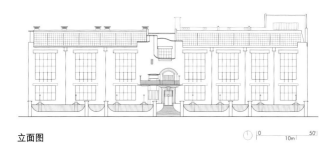

立面图

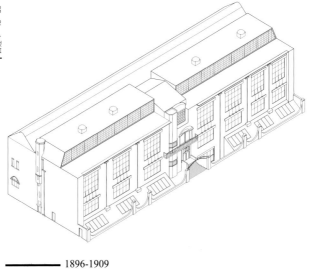

1896-1909

北纬 55°51′58″

GLASGOW SCHOOL OF ART
CHARLES RENNIE MACKINTOSH
Glasgow, Scotland

36

格拉斯哥艺术学校

查尔斯·伦尼·麦金托什

苏格兰，格拉斯哥

1896—1909 年

格拉斯哥艺术学校常常被看作欧洲第一批真正意义上的"现代"建筑，虽然是一所学校，它的外形却更像是一座工业厂房。这座砖石结构的建筑分两阶段建成，以其巨大的工业化生产的玻璃幕墙而闻名：它的玻璃窗时而舒展地嵌入砖石划分出立面节奏，时而变化为外凸的体块（凸窗）点缀陡峭的山墙。建筑平面呈轴线对称，按照"E"形布局组织内部的空间，其平面虽然并不复杂，室内的立体木质网格却增强了垂直方向上的空间连续性。为了避免日晒，面积较大的工作间被布置在北面沿街的一侧，光线透过宽阔的玻璃幕墙照亮了整个工作空间。建筑中大量使用了锻铁材料，有些是结构部件，有些则作为装饰元素出现——在一些位置，建筑师甚至刻意模糊二者的界限，例如精致的窗框和入口的拱门，它们既是结构也是装饰。这表明麦金托什已经实现了现代主义关于简化建筑材料的主张，将功能与形式合为一体。2014 年的火灾对建筑造成了很大损害，截至 2017 年修缮工作仍在进行。

首层平面图

立面图

1938-1939

北纬 61° 35′51″

VILLA MAIREA
ALVAR AALTO
Noormarkku, Finland

37

玛丽亚别墅
阿尔瓦·阿尔托
芬兰，诺尔马库
1938—1939 年

　　玛丽亚别墅代表了阿尔托创作生涯中一个重要的积累时期，他将早期的古典主义情怀与乡野的浪漫完全融入功能主义中，整栋建筑焕发着舒适自得的自然体验。这栋别墅的方形体块中糅合着蜿蜒的曲线，现代材料和技术也与传统材料工艺无缝对接，既富有对比，也显得相得益彰。由于别墅的主人痴迷于收藏现代艺术作品，阿尔托希望从现代绘画中的形式和意象中抽取要素，与建筑产生共鸣。别墅坐落在松涛叠叠的山顶上，阿尔托通过一系列手法将屋外的树林与空地揽入室内，建筑本身并不表现典型的节奏与韵律。结构部件与隔断屏风都显得纤细灵巧而不拘一格，建筑师称其为"模拟的"森林，呼应着房屋外面的林海。起居生活空间在房屋的形体中流动，各部分通过可移动的隔断划分出界限；几面墙体甚至都是可移动的，这样建筑就可以更加彻底地与自然融为一体。

东经 21° 52′28″

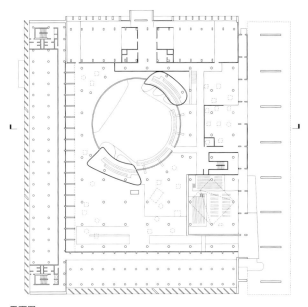

平面图

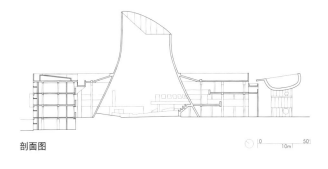

剖面图

北纬 30°45'40"

THE ASSEMBLY, CHANDIGARH
LE CORBUSIER
Chandigarh, India

38

昌迪加尔议会大厦

勒·柯布西耶

印度，昌迪加尔

1951—1964 年

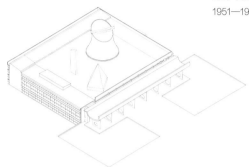

在柯布西耶为印度北部的旁遮普邦所作的新城规划及行政建筑群中，昌迪加尔议会大厦是最受瞩目的一项作品。这栋建筑共有四层，以稳定的"U"形布局围绕着三座雕塑般的巨型几何体。建筑入口处设置了一面宽阔的遮阳柱廊，顶端的巨型反拱屹然挺立，似乎是要擎住天空。建筑师以此形象提示出"张开的手"的建筑母题，这一符号在昌迪加尔建筑群中曾多次出现。议会大厦位于新城规划的中心位置，采用双曲面壳体混凝土结构（整个建筑群都采用了混凝土现浇的施工方法），四周的柱廊沐浴在漫射的自然光下，显得尤为动人。柯布西耶还亲自设计出一系列装饰元素来描绘旁遮普邦多元的文化风采：高达数层的彩色壁画动人心魄，巨门两侧的珐琅金属板和公共空间中用以点缀的挂毯，皆给人留下了深刻的印象。

1951-1964

东经 76°48'10"

88 /

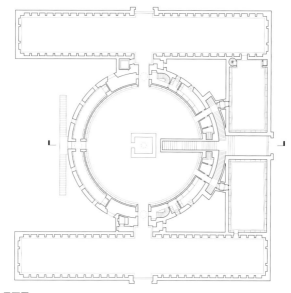

平面图

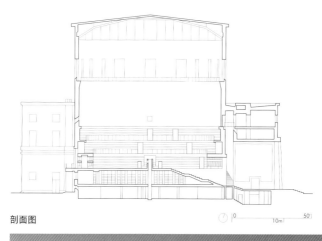

剖面图

1918-1927

STOCKHOLM PUBLIC LIBRARY
ERIK GUNNAR ASPLUND
Stockholm, Sweden

斯德哥尔摩公共图书馆
埃里克·贡纳尔·阿斯普隆德

瑞典，斯德哥尔摩
1918—1927 年

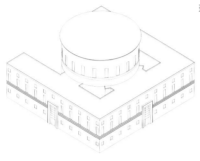

　　斯德哥尔摩公共图书馆可看作斯堪的纳维亚现代主义演进过程的缩影。阿斯普隆德起初想将它设计成一座新古典主义的建筑，创作过程中却逐步转向对于北欧古典主义几何抽象形式的探索，建筑最终落成后又被当成理性功能主义的典范。受到勒杜[1]所作的拉维莱特圆厅城关（Barrière de la Villette）的启发，建筑师将一座巨大的赭石色圆柱体放置在五层高的裙楼上面（裙楼的基座就占据了一层）。图书馆的布局精确地考虑了周边的环境，退居场地中央的大厅避让着两旁的林荫大道并强调出裙楼的转角，同时阿斯普隆德还根据视线做出微调，将建筑平面扭转了 4.5° 中心形体的后移也塑造了入口空间的形式：沿着修长而层叠的入口阶梯向上漫步，就到达了圆柱形的中央大厅，其高度是外露鼓座的两倍。中央大厅下部的三层布置了书架，大厅上空凹凸不平的墙体包裹的巨大中庭空间犹如天体一般宏伟壮观。

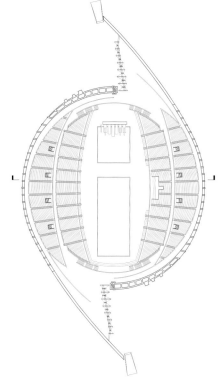

平面图

剖面图

NATIONAL OLYMPIC GYMNASIUM
KENZO TANGE
Tokyo, Japan

北纬 35°40′03″

40

代代木国立综合体育馆
丹下健三
日本，东京
1961—1964 年

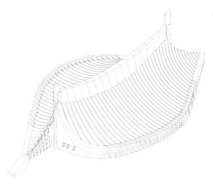

  代代木国立综合体育馆是为了迎接 1964 年东京奥运会而建造的，是当时全世界最大的悬索结构建筑。该项目一共包括两座场馆（规模较大的是游泳馆，较小的是篮球馆），抒情诗般的建筑形式与创意十足的结构方案融为一体，高超的手法令人称道。大体育馆的主体结构由两道几乎水平的巨拱构成，出挑的巨拱下方布置着建筑的入口和前厅，场馆内部空间可以容纳 10000 名观众。错置巨拱的端头浇筑了两部极高的混凝土塔楗，两根直径达到 13 英寸（约 0.33 米）的中央钢缆横贯其间形成了建筑的屋脊，支撑屋面板材的悬索则被连接在中央钢缆和巨拱之间。代代木体育馆代表了一种鲜明的日本现代主义风格，证明现代主义建筑即便摆脱具象的传统形式与地方性的建造技术，仍然具备丰富的文化包容性。

东经 139°42′00″

— 1961-1964

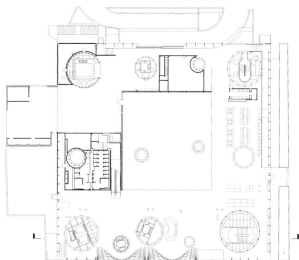

首层平面图

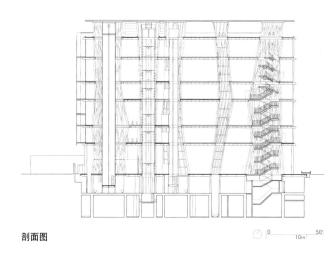

剖面图

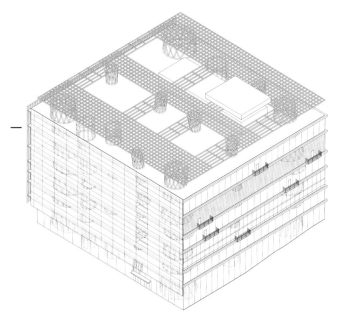

北纬 38°15′56″

41

SENDAI MEDIATHEQUE
TOYO ITO (ITO & ASSOCIATES, ARCHITECTS)
Sendai, Japan

仙台媒体中心
伊东丰雄（伊东丰雄建筑设计事务所）

日本，仙台
1995—2000 年

　　伊东丰雄与结构工程师佐佐木睦朗（Matsuro Sasaki）合作设计的仙台媒体中心采用了一套全新的结构体系。建筑的每一层空间都展现出极大的开放性和灵活性，与此同时却在视觉上反映出一种"无结构"的悖论。由钢结构组成的十三组巨大网状管道纵贯整栋建筑，如同扭转生长的树干一般支撑起层高不同的各层金属楼板。网状管道分为三列依次排布，既增加了结构的稳定性，同时也作为建筑内部的采光井；管道的大小与位置各不相同，承担着多种建筑功能：直径较小的管道是建筑的支撑结构，较大的管道内布置了电梯或楼梯，一些更小、更扭曲的"树干"里则包裹着电气和空气处理系统。在屋顶，位于建筑平面中心的两组管道上方设置了由计算机控制的反射镜，为建筑内部引入了自然光线。自由的平面布局与变化多端的剖面设计导致仙台媒体中心的四个立面得以根据功能产生不同的设计，同时由于建筑转角不设结构，极大地增强了立面的透明感与自主性。

东经 140°51′56″

1995-2000

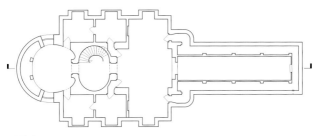

平面图

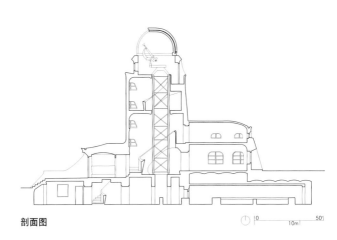

剖面图

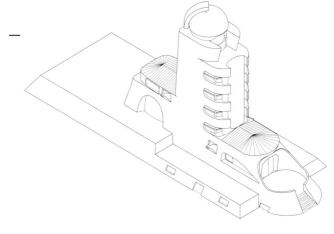

1917-1921

EINSTEIN TOWER
ERICH MENDELSOHN
Potsdam, Germany

42

爱因斯坦天文台
埃里克·门德尔松
德国，波茨坦
1917—1921 年

　　第一次世界大战期间，埃里克·门德尔松应征入伍。虽然战场上危机四伏，他却在战壕中抽空画出了许多小草图，并急切地想要将它们变成一种新型建筑。战争一结束，门德尔松就立刻投入了工作并在波茨坦建造了爱因斯坦天文台。他的草图中蕴含着一种"流动性"——既有些康定斯基[1]的表现主义（Expressionism）意味，还有点儿博乔尼[2]未来主义（Futurism）的风格。作为对建筑塑形能力的早期探索，门德尔松草图中的抽象元素被转化成爱因斯坦天文台独特的建筑形式，虚与实在流动的形体中彼此相映。最终，这座天马行空的建筑被恰如其分地献给了阿尔伯特·爱因斯坦。天文台本来计划使用钢筋混凝土结构，却因技术局限和材料匮乏而最终使用了砖砌的方法。爱因斯坦天文台中流露着高迪（Gaudí）早期作品中的优雅品味，雕塑般的体量感也在柯布西耶之后所作的朗香教堂中得到了再现。

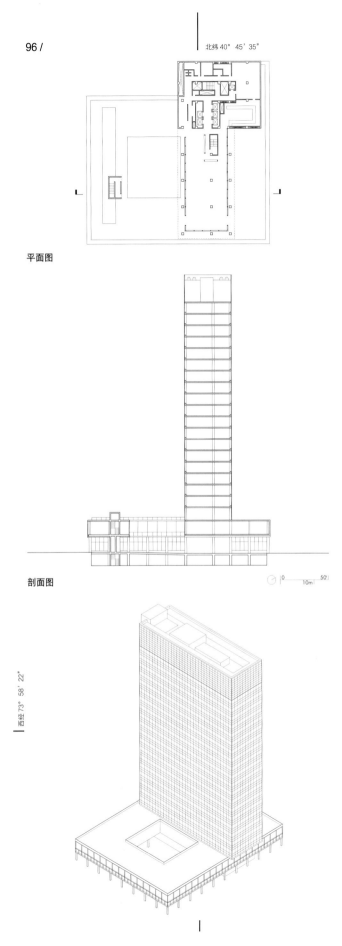

北纬 40°45'35"

平面图

剖面图

西经 73°58'22"

43

LEVER HOUSE
GORDON BUNSHAFT and NATALIE DE BLOIS (SOM)
New York, New York, USA

利华大厦
戈登·邦夏与纳塔莉·德布卢瓦（SOM 建筑设计事务所）

美国，纽约州，纽约
1951—1952 年

  利华大厦坐落于纽约市公园大道，是最早用于服务美国商业办公空间的现代建筑之一。它高耸的形体拔地而起，释放出大部分场地空间，阳光照亮街道，行人在建筑底层开敞的柱廊和花园中自由穿行。铝合金和玻璃构成的塔楼中包含了大部分的办公空间，大楼主体垂直于公园大道，交通核心筒被隐藏在建筑窄边远离街道的一侧。利华大厦是美国首批采用玻璃幕墙的商业建筑，它是戈登·邦夏留给纽约的一笔宝贵财富，同时也建立了 SOM（Skidmore, Owings & Merrill）在美国建筑界的崇高声誉。建筑落成后不久，密斯·凡·德·罗就在街道对面的南侧位置修建了西格拉姆大厦——这两栋不可思议的建筑拉开了现代建筑统治美国大城市的帷幕。

1951-1952

98 /

北纬 34°01'21"

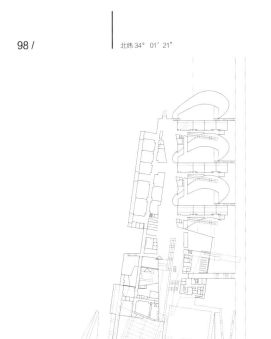

平面图

剖面图

西经 117°46'42"

DIAMOND RANCH HIGH SCHOOL
MORPHOSIS ARCHITECTS
Pomona, California, USA

44

钻石农场中学
墨菲西斯事务所
美国,加利福尼亚州,波莫纳
1994—1999 年

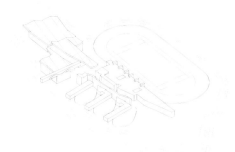

　　钻石农场中学坐落在一片开阔的场地中,基地一侧是陡峭的山崖,另一边则被一条高速公路切断。这片校园远远望去如同一座山城,模糊了建筑与景观之间的界限。建筑师参照地形学的规划方法应对场地中崎岖的斜坡,将各项功能嵌入山坡之中,并以线性组织方式颠覆了学校建筑的传统范型。沿着学校内的步行街行进,道路两侧是教学楼,路的尽头则是体育馆和自主餐厅。一系列线性建筑元素由内而外延伸,直抵斜坡下方的运动场。钻石农场中学是在加州标准公立学校基金的资助下建立的,学校的宣传口号似乎恰如其分地阐明了这座建筑:本校的建筑与本校课程一样优秀,皆能教化在此学习的学生并帮助他们成长。

1994-1999

北纬 41° 08' 33"

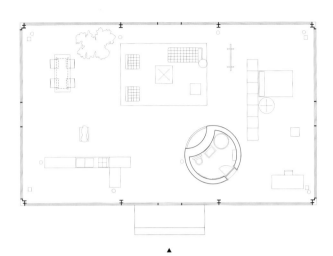

平面图

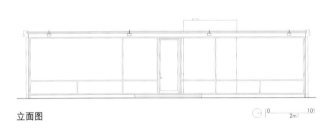

立面图

1949

GLASS HOUSE
PHILIP JOHNSON
New Canaan, Connecticut, USA

45

玻璃屋
菲利普·约翰逊
美国，康涅狄格州，新迦南
1949 年

　　虽然玻璃屋明显受到了密斯设计的范斯沃斯住宅的影响（密斯的方案成型更早，但两项作品几乎是同时建造），这座示范性的微型"玻璃住宅"却从截然不同的方向展开了创作：它的基座并不悬空，而是稳固地坐落在地面之上；柱子安装在玻璃幕墙之内而非表皮之外；建筑着重强调了形体的转角而非使其虚化；房屋内部也没有方形的"功能盒子"，而是设置了一个圆柱形的砖筒；比起范斯沃斯住宅所强调的透明性，玻璃屋更注重表现玻璃表面的反射；它还以黑色的钢结构与范斯沃斯住宅的白色结构形成对比，并以砖材铺地区别于范斯沃斯住宅的浅色大理石地面——可以看出，玻璃屋并不是一项"仿品"，它与范斯沃斯住宅产生了一系列辩证的比照，约翰逊将密斯的建筑宣言转译成了一座实实在在的美国建筑。这两栋别墅都依仗业主雄厚的财力来为居住者提供隐私空间[1]，但约翰逊在建筑周围设置了低矮的景观墙，还在不远处修建了一间独立的小房子，以补偿玻璃屋"简朴的"建筑功能。

北纬 40°44'60"

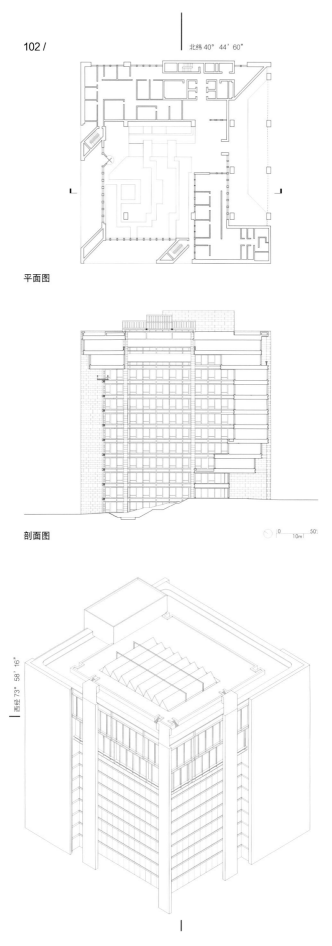

平面图

剖面图

西经 73°58'16"

FORD FOUNDATION HEADQUARTERS
KEVIN ROCHE, JOHN DINKELOO and ASSOCIATES, LLCE
New York, New York, USA

46

福特基金会总部大楼

凯文·洛奇与约翰·丁克洛建筑事务所

美国,纽约州,纽约

1963—1968 年

福特基金会总部大楼位于纽约曼哈顿中心区,是一栋构思奇特的办公楼。该建筑地上部分共有 11 层,其中 1 至 9 层的办公空间呈 "L" 形半包围布局,跨度较窄,10 层和 11 层则呈闭合而中空的 "O" 形布局。这 11 个楼层包围着一个巨大的通高中庭,建筑师在中庭底层设置了一片宽阔的室内花园。阳光通过建筑东南面巨大的玻璃幕墙和上方的桁架屋顶照亮了中庭,底层的花园显得生气勃勃。建筑主体结构采用了耐腐蚀的戈坦钢(Cor-Ten steel)。立面则采用石材和玻璃,与钢材形成平衡。虽然东立面与南立面的外部造型近似,却展现了不同的空间深度:透过中庭外的玻璃幕墙,内部的空间与形体清晰可见。主入口位于建筑北侧,门厅顶端的 3 个楼层依次退进,这不但协调了建筑体量和场地的视觉关系,还表现出"欲扬先抑"的设计手法:当人们穿过厚重的悬挑入口,终于看到了内部明亮而宽阔的中庭和花园,豁然开朗的感受油然而生。

1963-1968

104 /

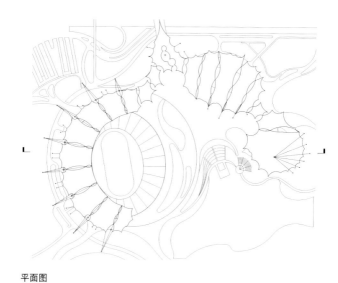

平面图

剖面图

北纬 48°10′24″

OLYMPIC STADIUM, MUNICH
FREI OTTO and GÜNTER BEHNISCH
Munich, Germany

47

慕尼黑奥林匹克体育场
弗雷·奥托与金特·贝尼施
德国，慕尼黑
1968—1972 年

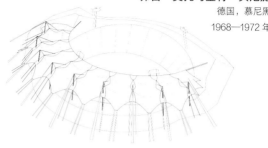

　　1972 年第 20 届夏季奥运会在德国慕尼黑举行，这是二战后奥运会第一次在德国本土举办（纳粹德国曾在 1936 年举办柏林奥运会）。与前几届奥运会厚重而富有纪念性的场馆截然不同——在慕尼黑奥林匹克体育场的设计中，奥托与贝尼施将一片轻薄的网索屋面覆在连续不断的看台上，令建筑与周围壮丽的地景融为一体。虽然搭帐篷是最为古老的结构做法之一，慕尼黑奥林匹克体育场中的网索结构屋面却使用了当时世界上最复杂的材料和设计技术。建筑使用管状钢柱作为支撑结构支起主钢缆，主缆上连接的较细钢索交织成一张巨大的网索，网索之间安装着涂有 PVS（聚乙烯硫酸）的聚酯胶板或丙烯酸玻璃板。整体结构首先经由计算机建模演算，确定无误后在现场施工安装。山峰般的网索屋面从风景如画的场地中浮起，形式极其优美动人（建筑师以此形式隐喻了附近阿尔卑斯山下的风光）。慕尼黑奥林匹克体育场的基址也值得一提：那片凹陷的场地是第二次世界大战时期遭到轰炸而形成的一个巨大弹坑。

1968-1972

东经 11°32′48″

106 / 北纬 34°02′07″

首层平面图

立面图

西经 118°29′04″

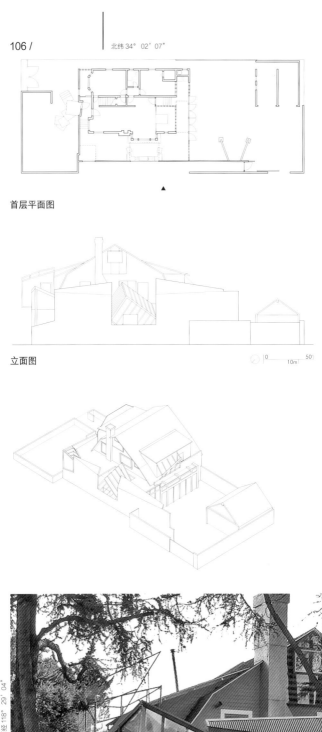

GEHRY HOUSE
FRANK GEHRY (GEHRY PARTNERS, LLP)
Santa Monica, California, USA

48

盖里自宅
弗兰克·盖里(盖里建筑事务所)
美国,加利福尼亚州,圣莫尼卡
1979 年

  盖里对自己荷兰殖民风格(Dutch Colonial)的石棉板小屋进行了大刀阔斧的改建,三个建筑立面的外观和形体都发生了剧烈变化:波纹金属板、钢丝围网、原始胶合板全都派上了用场,门窗也都经过重新设计,房屋原本的形式被这些新元素侵蚀殆尽。改建后,各式各样的材质相互映照,密密层层地叠盖在建筑中,这些材料有时候直接表达了室内空间,有时却又发生位移,表现出错置的关系。盖里自宅中材料的排列组合令人想起了劳申贝格[1]的拼贴画,形体的错位和重置则令人联想到杜尚[2]的"现成品艺术",这项作品可说是站在了建筑和艺术品的中间位置。盖里自宅充满实验性与不确定性,立场非常主观;与此同时,其中的轻捷木骨架、彩绘窗框和拼贴线脚也表现出怀旧的情愫。盖里对于各种材料的运用可说是驾轻就熟,因此这栋建筑的形式中体现了一种美国化的工业感。盖里自宅以一种"开放的混乱",重释了20 世纪建筑整洁的层次结构、方法论和宏伟目标。

— 1979

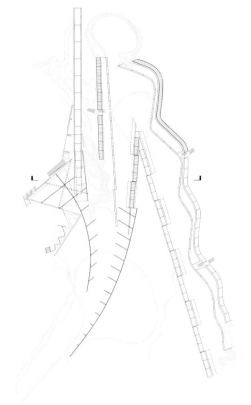

场地平面图

剖面图

IGUALADA CEMETERY
ENRIC MIRALLES and CARME PINÓS
Barcelona, Spain

49

伊瓜拉达墓园

恩里克·米拉莱斯与卡梅·皮诺斯

西班牙，巴塞罗那

1985—1991 年

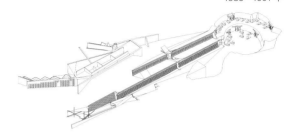

  与历史上所有的伟大墓葬建筑一样，伊瓜拉达墓园游走于隐喻与神话之间。墓园位于郊外的工业区，开采和挖掘已经让这片土地沟壑遍布，为了使访客忽略周围复杂的环境，建筑师将墓园安排在一处低洼的地形中。与此同时，该作品大部分使用钢筋混凝土为材料，暴露着钢筋的表皮如顽石迸裂，露出的钢筋本身则像是空气中凭空绘出的线条，雕凿着建筑的轮廓。伊瓜拉达墓园中的建筑元素十分朴素：石笼墙（gabion walls）上的钢网已经锈迹斑斑，废弃的铁轨枕木被嵌入粗糙的水泥地面中——都是一些人们熟悉的形象。一条路径从园中穿过，其一端连接着场地底部的空谷，沿着脚下的枕木向上方游走，访客的视线被引向头顶的苍茫天空。路径最终停歇在墓园的椭圆形广场旁，几处倾斜着横躺的盖板隐喻着掘出的墓穴，寓意逝者将在此安息长眠。

1985-1991

北纬 41°49'60"

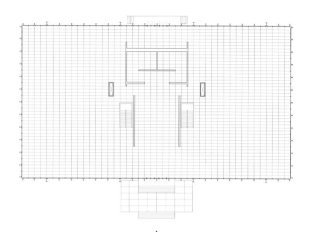

首层平面图

立面图

CROWN HALL
MIES VAN DER ROHE
Chicago, Illinois, USA

50

克朗楼
密斯·凡·德·罗
美国，伊利诺伊州，芝加哥
1950—1956 年

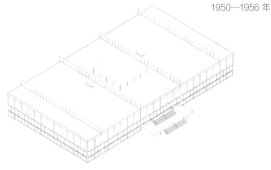

　　克朗楼是密斯为芝加哥的伊利诺伊理工大学（Illinois Institute of Technology）设计的建筑系馆，展现出密斯·凡·德·罗二十世纪五十年代的典型作品风格：对于纪念性的表达以及中性的功能规划。这栋建筑乍看之下只是一个普通的玻璃盒子，实际上其细节和比例都极为考究。晶莹的玻璃形体被轻轻托举，庙宇一般矗立在基座之上，水平屋顶则被坚固的桁架轻轻提起。建筑平面基本对称，按照比例关系布置出各部分的尺度和位置，主要功能都布置在边沿，较为封闭的服务空间则被放置在建筑的中心位置。这栋简洁的建筑阐述了一个与功能相关的理想主义建筑愿景：它既不针对一个特别的时间，也不拘于一处特定的地点，却包含着通用性的整体建筑潜力，也记录了密斯对建筑纪念性的执着表现。

1950-1956

北纬 40°04'14"

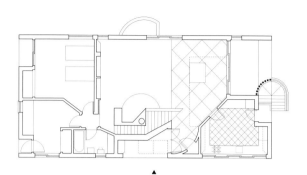

首层平面图

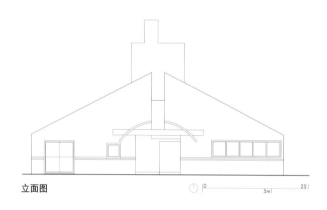

立面图　　0　5m　25'

西经 75°12'29"

VANNA VENTURI HOUSE
ROBERT VENTURI and DENISE SCOTT BROWN
Chestnut Hill, Pennsylvania, USA

51

母亲住宅
罗伯特·文丘里与丹尼丝·斯科特·布朗
美国，宾夕法尼亚州，栗树山
1962—1964 年

母亲住宅是后现代主义建筑（postmodernism）的开山之作。它的立面形式吸取了勒·柯布西耶的斯坦别墅（Villa Stein）与路易吉·莫雷蒂[1]的太阳花住宅（Casa Girasole）的特征，平面布置中则流露出勒琴斯[2]与麦金[3]的风格。这栋建筑师为其母亲设计的居所既是对现代主义的致敬，也是对其发起的挑战。各个建筑元素时常发生分裂和位移：屋内的壁炉争夺着中心地位，挤压了楼梯；二层空间的体量与厨房的对角线一道，侵入横跨在餐厅上方的拱形空间；在建筑入口处，一条断裂的拱门线脚拼贴在门梁上方，制造出分裂的视觉感受；建筑正立面在形状和体量上完全对称，却又安插着不同形式的窗户；在建筑背面，笔直的单坡屋顶下却挑出了一面弧形的阳台窗，感觉甚至有些"别扭"。这项作品的轮廓非常平淡，试图伪装成平平无奇的样子，而各处的细节却又抓住人们的眼球引起注意。文丘里将"建筑的复杂性和矛盾性"的论调播撒进这栋住宅中，同时也塑造了一处充满纪念性的家庭生活场景。

1962-1964

114 /

北纬 34°07′05″

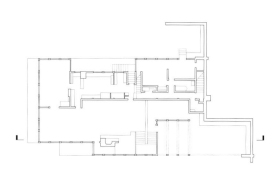

平面图

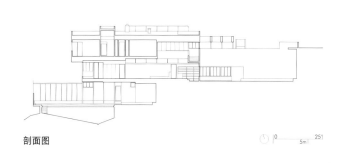

剖面图

西经 118°17′16″

1927-1929

LOVELL HEALTH HOUSE
RICHARD NEUTRA
Los Angeles, California, USA

52

洛弗尔健康之家
里夏德·诺伊特拉
美国,加利福尼亚州,洛杉矶
1927—1929 年

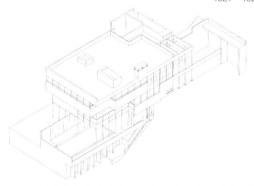

  随着洛弗尔健康之家那令人震惊的轻薄结构被锚固在山体之上,洛杉矶便诞生了一种全新的建筑形式。这栋坐落在市郊的山地别墅不但是当时首批使用钢结构建造的住宅,还开创了喷涂混凝土的施工工艺,即不用模具浇筑,直接将混凝土喷射在保温墙板上面(也就是现在的喷射混凝土)。这栋通透明亮的建筑还是一座名副其实的"健康之家",其主人是一位信奉自然疗法的理疗师,推崇裸体日光浴和室外睡眠,相信大自然的空气和充足的阳光能为人带来活力。正是基于业主独特的功能需求,诺伊特拉才得以在设计中尝试全新的建筑材料和技术。健康之家的阳台悬浮般出挑着,开放式的室内空间与外部的自然环境紧密联通;透过宽阔的落地窗,远方的洛杉矶城市景观尽收眼底。

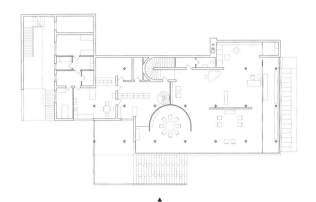

首层平面图

立面图

1928-1930

北纬 49°12′26″

TUGENDHAT HOUSE
MIES VAN DER ROHE
Brno, Czech Republic

53

图根哈特住宅
密斯·凡·德·罗
捷克，布尔诺
1928—1930 年

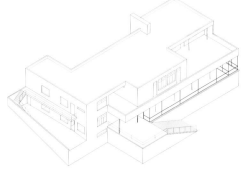

　　如果将巴塞罗那博览会德国馆天马行空的设计构思进行"驯化"，便得到了图根哈特住宅。在这栋住宅中，加厚的墙壁围合着卧室，乳白色玻璃也不再凌厉[1]，十字形钢柱的轮廓变得圆润起来。建筑共有 3 层，各个层面在山坡上穿插错落：入口出乎意料地设在建筑顶部，人进入后便被引向一部弧形楼梯；向下行进就进入了生活起居空间，光线十分明亮。起居室东边与南边都设置了巨大的落地玻璃幕墙，几乎横贯整个建筑立面，透过玻璃窗便可远望山坡下的布尔诺市。为了表现室内外空间的流动与转换，建筑二层的玻璃幕墙可以通过滑轨降入地下室内，让起居室变成一片开放的室外平台。一扇由望加锡黑檀木（Macassar ebony）制成的半圆屏风围起餐厅，主人可以在这个私人的景观平台上休憩，眺望远方的城市风光。

东经 16°36′58″

118 /

北纬 42° 21′ 25″

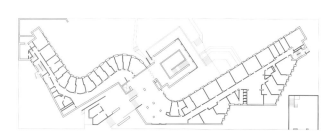

平面图

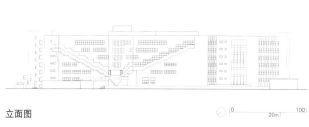

立面图

西经 71° 05′ 45″

1947-1948

BAKER HOUSE
ALVAR AALTO
Cambridge, Massachusetts, USA

54

贝克公寓
阿尔瓦·阿尔托
美国，马萨诸塞州，剑桥
1947—1948 年

  建筑师阿尔瓦·阿尔托一直特别注重室内环境与室外环境的融合。在担任麻省理工学院的客座教授时，他设计了该校的贝克公寓。在方案设计中，阿尔托非常重视场地附近的查尔斯河（Charles River）可能产生的影响，也明白阳光对于一栋位于马萨诸塞州的学生公寓的重要性。[1]他以内部走廊为界，将宿舍布置在建筑南侧，砖材砌筑的立面模仿着河流的形式，蜿蜒而曲折；北面则主要布置了公共空间和数部楼梯，立面是锯齿状的折线形态，混凝土和玻璃材质的使用也显示出更强的"城市建筑"特征。贝克公寓蜿蜒的体量延伸了立面的长度，所有宿舍都尽可能朝向阳面，宿舍的景观视野也得以开拓，能让学生在房间内眺望查尔斯河的风景，也使宿舍的窗户避开了北面嘈杂的城市公路。贝克公寓是学校宿舍建筑设计的经典范例，阿尔托将学生的私人空间与社交空间处理得井井有条，互不干扰却又紧密联系，其高超的手法受到了广泛赞誉。

120 /

北纬 19°24′39″

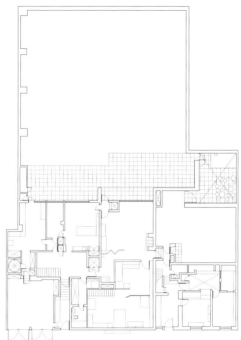

首层平面图

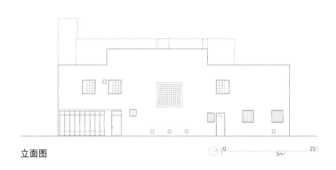

立面图

西经 99°11′31″

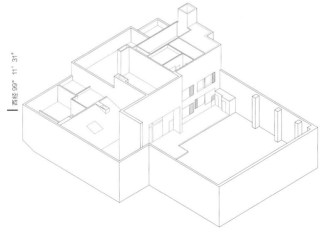

1947-1948

55

BARRAGÁN HOUSE AND STUDIO
LUIS BARRAGÁN
Mexico City, Mexico

巴拉甘自宅与工作室
路易斯·巴拉甘
墨西哥,墨西哥城
1947—1948 年

路易斯·巴拉甘是二战后新一代现代建筑师中的代表人物。这一批建筑师大多并非生于欧洲,设计观念直接受到本土文化和社会价值的启示和影响,创作手法丰富而多样,他们以多元化的作品挑战了现代建筑早期的功能主义(functionalist)学说与似是而非的理性主义(specious intellectualism)。巴拉甘自宅展现了强烈的地域性特征,建筑通过层层递进的空间组织,在兼顾功能的同时还在各个位置营造出截然不同的建筑体验。住宅位于街道旁边,形体并不显著,灰色的抹灰墙面上有几处看似随意布置的窗洞,看上去和周围的房屋一样封闭而平庸。一旦进入建筑内部,就会被推入一系列"去物质化的"(dematerialize)、色彩浓烈的抽象空间,幽密的花园若隐若现地从夹缝和窗户中显露出来,引人去寻求探访。从嘈杂的城市进入色彩鲜明的房间,再游入花园的景观,最后登上屋顶纯净的露台——整个空间转换过程明显经过了精密安排。身为一位墨西哥建筑师,巴拉甘在实践中承认并积极介入其所处的文化背景,描绘出一种别具一格的现代建筑风貌。

北纬 41° 18' 42"

平面图

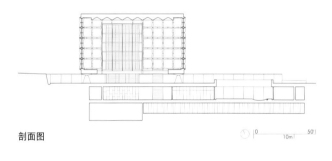

剖面图

THE BEINECKERARE BOOK & MANUSCRIPT LIBRARY
GORDON BUNSHAFT (SOM)
Yale University, New Haven, Connecticut, USA

拜内克古籍善本图书馆
戈登·邦夏（SOM 建筑设计事务所）

美国，康涅狄格州，纽黑文
1960—1963 年

耶鲁大学的拜内克图书馆是一座超凡的建筑，也是戈登·邦夏最具诗意的成熟作品。这座图书馆不仅是一座珍本书籍的宝库，更像是一处虔诚的圣所，由佛蒙特大理石板组成的白色体块安静地栖居在四座巨大的水泥柱墩上面，显出一种与世隔绝的超然姿态。[1] 邦夏从耶鲁大学历史悠久的校园中搜寻出建筑构造和材料的线索，图书馆立面轻薄的大理石板犹如一面面"石窗"：在白天，建筑被太阳照亮，室内洒满了斑驳而柔和的琥珀色光线；在夜里，内部的灯光渗出表皮，建筑如灯笼一样照亮了周围的小广场。图书馆内部矗立着一座由青铜和玻璃建造的"书塔"，极为精致。一座下沉的庭院嵌入建筑旁边，将光线引入藏于地下的手稿收藏室。庭院中放置着一组野口勇[2]创作的石景，也是大理石材质的，重复并且加强了建筑的主题。

1960-1963

124 /

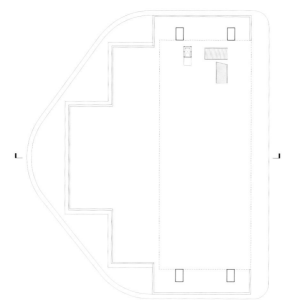

首层平面图

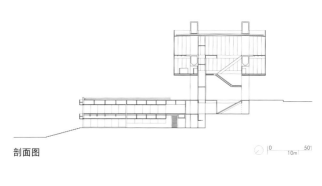

剖面图

MUSEU DE ARTE DE SÃO PAULO
LINA BO BARDI
São Paulo, Brazil

57

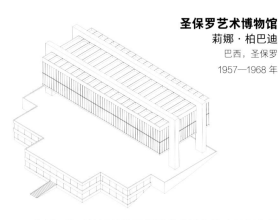

圣保罗艺术博物馆
莉娜·柏巴迪
巴西，圣保罗
1957—1968 年

　　莉娜·柏巴迪设计的圣保罗艺术博物馆坐落于特里亚农公园（Trianon Park）的边缘，紧邻圣保罗人大道（Avenida Paulista），倾斜的基地下方还有一条高速公路穿过，博物馆以精密的布局策略和结构方式应对了复杂的场地情况。四个巨型的混凝土支柱举起两组巨大的横梁，一组横梁提起屋顶，另一组则布置在建筑中部；一系列反梁连在横梁之间，支起屋面和楼板——从外面看过去，圣保罗艺术博物馆就像是一个悬挂在混凝土框架之下的水晶盒子。博物馆下方的露台是特里亚农公园的延伸，人们从公园出来后便可走上平台，俯瞰下方的城市。这片露台也将博物馆分成了上下两部分，其中各自安置了不同的功能。柏巴迪一贯强调建筑的公众属性，这项作品最初的设计意图便是建造一座向城市四面开放的巨型玻璃盒子，其余的功能（剧院、餐厅和服务用房）则退居其后，被埋入基座的地下空间中。

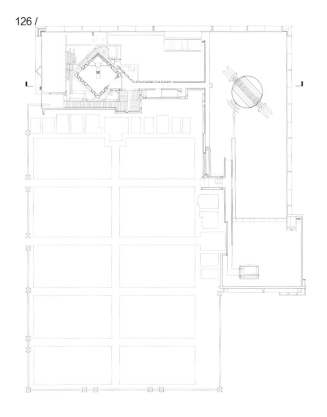

平面图

剖面图

北纬 45°45′05″

BRION FAMILY TOMB
CARLO SCARPA
Treviso, Italy

58

布里翁墓园
卡罗·斯卡帕
意大利,特雷维索
1969—1978 年

穿过圣维托村简朴的村庄墓地,就来到了布里翁墓园。它的规模和体量与附近的家族墓相仿,从远处看来并没有太大差异。穿过山门,便来到一条南北向的游廊,透过墙中央的双环洞口可以看到园内的景色(双环的元素在这项作品中反复出现),此时才发觉这片墓园绝不仅仅是为了引发人们对死亡的思考而建造的:这是一个分形的[1]世界,借由视线的不断延伸——转过角落,翻越围墙,直抵地平线——墓园的界限似乎在不断扩张,建筑的每一部分似乎都能够分裂出更多的局部。墓园呈"L"形布局,西南两侧被村庄公墓包围。按照由南向北的顺序,首先可以看到静思亭,它伸入了一片沉静的水面;其后是布里翁夫妇的陵墓,它位于"L"形的转角位置,夫妇二人的石棺被安放在一块圆形凹地上面,一跨宽阔的拱桥遮盖在棺椁上方;向西行进便到了家族墓地,再往前走就到达了葬礼教堂。整个过程一直重复出现关于大地与天空、水与光的建筑主题。斯卡帕在布里翁墓园中塑造出一种诗化的空间体验,沁人心脾的场所却更胜诗歌的词句。

东经 11°54′49″

1969-1978

平面图

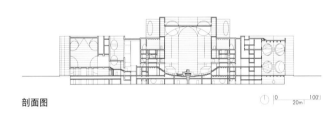

剖面图

59

北纬 23°45'46"

NATIONAL ASSEMBLY BUILDING, BANGLADESH
LOUIS KAHN
Dhaka, Bangladesh

孟加拉国国会大厦
路易斯·康

孟加拉国，达卡
1962—1983 年

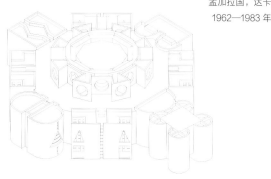

　　孟加拉国国会大厦是一座由砖块和钢筋混凝土建造的纪念碑，其纯粹的几何形式中记录着时间的流逝和太阳的运转。自然光从建筑中心的巨大天窗泻下，照亮中央大厅；建筑外围的体块环聚在大厅四周，包覆着议会的各项功能。从外看去，建筑形体基本上保持了一种封闭的态势，几个形式抽象的开口是为了将周围潟湖的冷空气吸入建筑内部。巨大的形体、不断变换的尺度、光与影的交替，建筑师以此刻画出超然的空间，偶尔出现的栏杆或楼梯又将空间拉回了人体尺度的范畴。孟加拉国国会大厦中没有一根柱子，反过来，也可以说整栋建筑都是由巨大的、可使用的中空巨柱构成的。[1] 路易斯·康为达卡所作的首都综合体规划也十分著名，它与卢伊藤斯（Luytens）的新德里规划、尼迈耶的巴西利亚规划、柯布西耶的昌迪加尔规划一道，都体现出发展中国家建立现代化首都的城市愿景。

东经 90°22'42"

1962-1983

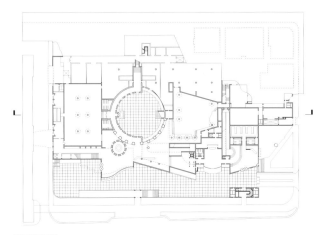

首层平面图

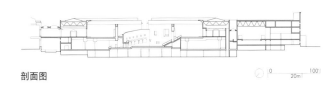

剖面图

NEUE STAATSGALERIE
JAMES STIRLING, MICHAEL WILFORD and ASSOCIATES
Stuttgart, Germany

60

斯图加特美术馆
詹姆斯·斯特林与迈克尔·威尔福德事务所

德国,斯图加特

1977—1983 年

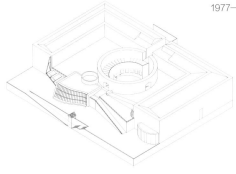

  在 1977 年的斯图加特美术馆竞标中,斯特林的方案力拔头筹。他将一系列紧密交织的广场与庭院布置在倾斜的场地中,模糊了博物馆与公共空间的界限。斯特林设计的新馆建在了历史悠久的老美术馆旁边,与其直接连通,新的方案遵照原有建筑"U"形的布局,将平面也设置为对称的形式。画廊部分沿着"U"形布局依次串联排列,并采用了顶部采光的方式。新馆的中心是一片圆形的露天庭院,建筑师围绕着庭院的边沿布置出公共交通流线。建筑的入口空间采用了较为活泼的设计元素,打破了平面井井有条的层次结构,流畅的曲面斜切入主要形体,似乎是在邀请着人们进入美术馆。斯特林在斯图加特美术馆的设计中实践了柯林·罗[1]在《拼贴城市》一书中提出的理念,以建筑调和了历史、都市生活、城市文脉、城市序列等概念,并在抽象和表现间取得了微妙的平衡。

1977-1983

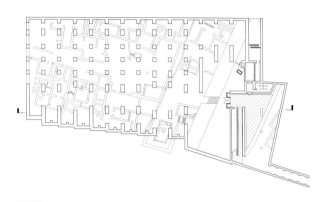

平面图

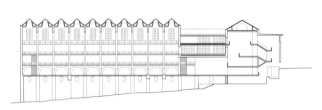

剖面图

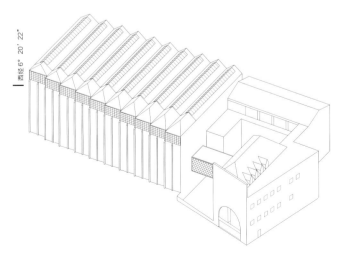

61

NATIONAL MUSEUM OF ROMAN ART
RAFAEL MONEO
Mérida, Spain

西班牙国家罗马艺术博物馆
拉斐尔·莫内奥
西班牙，梅里达
1980—1986 年

西班牙国家罗马艺术博物馆是一座现代的巴西利卡（Basilica），同时表现出历史性与现代性。这座建筑毗邻梅里达古罗马建筑群，莫内奥用古老的砌造技术在遗迹上方建起一系列高耸的平行砖拱墙，阳光穿过透明的顶棚，在拱券上映照出辉煌的光彩；拱墙下布置着一条由轻钢建造的回廊，它们不但将砖墙串联为一个整体，还使建筑脱离了对于古风的单纯模仿，引发了考古遗迹与当代城市环境的对话。博物馆的侧面是梅里达街（Calle Mélida），此处的建筑立面呼应了旁边古罗马剧场的风貌，二者的比例与造型都非常相似。建筑的结构网格是依照目前城市街区的脉络设置的，这与位于建筑地下层的古罗马遗迹的格局发生了冲突。新建的结构却没有侵扰遗迹的空间尺度，它轻柔地覆盖在历史之上，将古罗马的废墟变成展示内容的一部分，等待人们前来鉴赏。

1980-1986

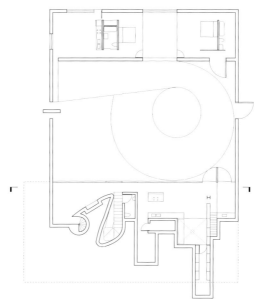

首层平面图

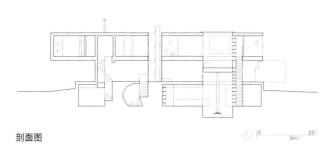

剖面图

北纬 44° 49′ 42″

MAISON À BORDEAUX
REM KOOLHAAS (OMA)
Bordeaux, France

62

波尔多住宅
雷姆·库哈斯(大都会建筑事务所)
法国,波尔多
1994—1998 年

  波尔多住宅的主人曾遭遇车祸,只能依靠轮椅行动,这栋建筑便是为他"量身定制"的。建筑师将三个"房子"叠盖起来,用一部大型升降平台将各层"串在一起",每一层既独立又相互联系。在升降平台的一侧布置了一面贯通三层的"书架墙",上面容纳了男主人的"一切生活所需",书籍、艺术品和红酒一应俱全。波尔多住宅以这部升降梯作为设计的核心,业主可以通过升降平台到达各层,生活起居毫无障碍;在平面中对称的位置,建筑师为其他家庭成员设置了一部螺旋楼梯。这栋建筑坐落在波尔多郊外的一处山坡上,顶层是一个巨大的混凝土方盒子,依照视线设置的一系列圆形舷窗起到了"景框"的作用,透过它们可以看到远处城市的风光;中间层被落地玻璃幕墙包裹着,显得极为轻薄,其透明表皮不但暴露出建筑内部的结构,也与封闭的顶层形成对比,加强了顶层体块的"漂浮感"。

1994-1998

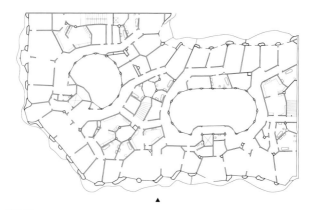

平面图

立面图

1906-1912

北纬 41°23'44"

CASA MILÀ
ANTONI GAUDÍ
Barcelona, Spain

米拉公寓
安东尼·高迪
西班牙,巴塞罗那
1906—1912 年

  米拉公寓的造型看似随意,却基于理性的规划和结构原则。这栋建筑并不是简单地将承重墙砌筑成"有机形式",而是采取了更为复杂的结构方案,将柱子、拱券和支撑钢梁沿着立面的曲线严谨地组织在一起,雕刻出蜿蜒曲折的建筑造型。建筑表皮的石灰石被仔细切割出特定的曲线,连石板间的交接线也做成了不规则的线条[这栋建筑被当地人称为"采石场"(The Quarry)便是由此而来]。建筑外部的铁艺构件精心设计为植物和骨骼的形式,十分精巧。米拉公寓的形式与表现手法大大改变了人们对"直线条建筑"的常规认识,也奠定了"加泰罗尼亚新艺术风格"(Catalan Art Nouveau)的基本特征。屋顶上还设置着一个宽阔的阶梯形景观平台,高耸的烟囱体现出强烈的纪念性,天窗和楼梯通道也别具一格;建筑中心设置着两个陡峭的中庭,阳光泻入其中,照亮了下层的庭院。

东经 2°09'43"

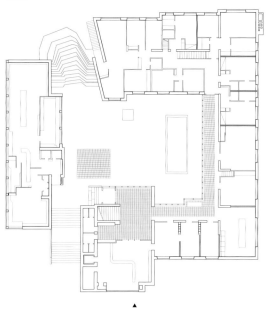

平面图

立面图

北纬 62°08′25″

SÄYNÄTSALO TOWN HALL
ALVAR AALTO
Säynätsalo, Finland

64

赛于奈察洛市政厅
阿尔瓦·阿尔托
芬兰，赛于奈察洛
1949—1952 年

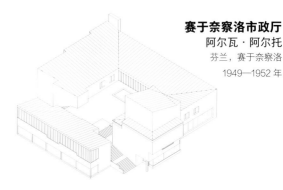

　　赛于奈察洛市位于芬兰中部的一座岛屿上，是一座拥有 3000 名居民的新兴工业小城。阿尔托赢得了该市市政中心的竞标，其平面布局受到文艺复兴时期庭院的启示，明显令人联想到传统的欧洲城市广场（阿尔托曾在一份笔记中表示自己受锡耶纳坎波广场的启发设计了该方案）。赛于奈察洛市政厅采用砖木结构，"C"形的建筑布局围合着中心的庭院，连续转折的体量以会议厅的砖塔为终结。位于庭院南端的体块与建筑主体分离，其上层是一个小图书馆，下层则是商铺。建筑平面虽然参照了意大利城市广场的经典布局，立面却展现出现代感，从构造的细部处理到窗户、门洞的定位，再到室外景观环境，市政厅的建筑形象体现出丰富的层次关系；西侧宽阔的台阶上以层层的青草覆盖了生硬的踏步，令人印象深刻。阿尔托擅长用当地材料与构造技术营造建筑，再加上他对类型学的抽象解读，使其成为同时代中最具地域性特色的现代建筑师。

东经 25°46′09″

1949-1952

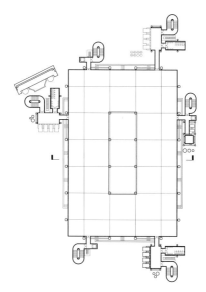

平面图

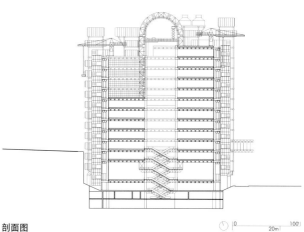

剖面图

西经 0°04'56"

北纬 51° 30′ 48″

LLOYD'S OF LONDON
RICHARD ROGERS
London, United Kingdom

伦敦劳埃德大厦
理查德·罗杰斯
英国,伦敦
1978—1986 年

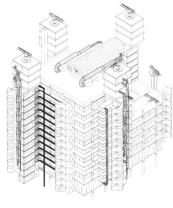

与蓬皮杜中心一样,理查德·罗杰斯设计的伦敦劳埃德大厦也是一座"由内而外"的建筑,不过这一次的项目是一座高层塔楼。劳埃德大厦的电梯、楼梯、管道系统、电气管网、设备间甚至洗手间等功能构件全部暴露在外,建筑表皮上则覆盖着不锈钢面板和玻璃幕墙。可以说,劳埃德大厦以自己"物化"(materiality)的设计风格为遍布历史建筑的伦敦金融区带来了高技派(high-tech)建筑美学。罗杰斯在建筑中心布置了一个高达 60 米的中庭空间,打开了楼层内的视野,他还将建筑顶层用于施工的起重塔吊保留下来,以便未来对建筑进行加建。这些塔吊如同"高技派滴水兽"(techno-gargoyles)般悬立于高空,此种情形对于这栋"机器建筑"而言再合适不过了。

1978-1986

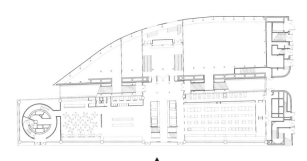

平面图

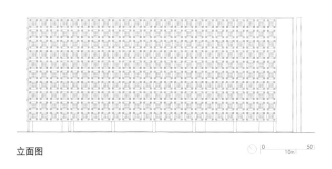

立面图

INSTITUT DU MONDE ARABE
JEAN NOUVEL, LÉZÉNÈS, PIERRE SORIA,
ARCHITECTURE – STUDIO
Paris, France

66

阿拉伯世界研究中心

让·努韦尔，吉尔贝·莱泽涅斯，皮埃尔·索里亚与 AS 建筑工作室

法国，巴黎

1981—1987 年

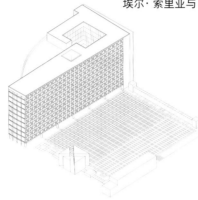

　　阿拉伯世界研究中心是对建筑场地与项目性质进行多重辩证的结果。从西侧望去，它似乎是两栋独立的高层建筑，但是二者又借由一个立方体般的中庭结为一体，令人联想到阿拉伯建筑内向型的空间特征[1]；建筑的形体看似极为简单，却又包含着非常复杂的装饰细节。南立面内的动态金属遮阳窗是这栋建筑最显著的特征之一，这项装置可以主动调节室内的光环境，感光元件控制的金属窗孔能够根据外部光线情况控制遮光叶片光圈的大小，在室内营造出斑驳的光影效果，与阿拉伯传统建筑中的木格雕画窗（mashrabiya）有异曲同工之妙。阿拉伯世界文化中心将新技术与传统建筑形式融合，在阿拉伯文化与西方文化之间架起了一座桥梁。

1981-1987

场地平面图

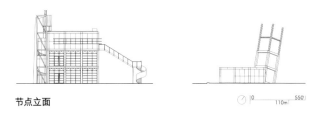

节点立面

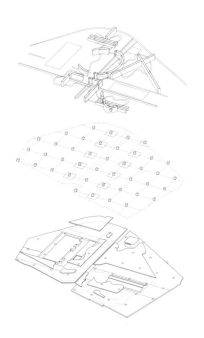

—

PARC DE LA VILLETTE
BERNARD TSCHUMI
Paris, France

67

拉维莱特公园
伯纳德·屈米
法国,巴黎
1982—1998 年

拉维莱特公园位于一片开阔的场地上,中部有一条运河横穿而过,基地中还有一座大型博物馆及一片巴黎肉类市场的遗留建筑;如此复杂的场地环境为所有参加公园方案竞标的建筑师出了一道难题。伯纳德·屈米的设计主题是"建造一座21世纪的公园",他绕开传统的"园林公园"形式,试图营造出一片充满互动、探索和再发现的场所。屈米在平面中叠盖了数层复杂线条,并以弯曲的路径和几何化的图形将不同的意义融合在一起;其后,他根据场地内现存的屠宰场的几何结构发展出一个网格系统,并在网格的交点上布置了一系列被称为"疯狂物"(follies)的建筑装置。这些装置被漆成鲜红色,表现出新构成主义(neo-Constructivist)的风格特征,可根据日后的需求拆除或者加建。它们同时也作为视觉参照物,提示出公园各部分之间的距离与尺度。可以说,拉维莱特公园是抽象形式与设计理性相互叠盖的结果,证明了一个抽象的设计方案也有被公众理解并充分使用的可能性。

146 /

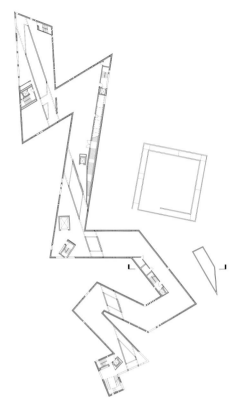

平面图

剖面图

北纬 52°30′08″

JEWISH MUSEUM BERLIN
DANIEL LIBESKIND (STUDIO LIBESKIND)
Berlin, Germany

68

柏林犹太人博物馆
丹尼尔·里伯斯金（里伯斯金工作室）
德国，柏林
1989—1999 年

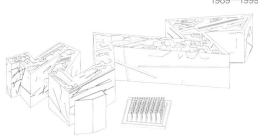

　　柏林犹太人博物馆刻意塑造出一种冷峻而疏离的建筑体验，这与博物馆主办机构的博爱与包容形成一种鲜明的对比。这栋建筑充分体现了里伯斯金在多维度上叠加信息的设计观念。建筑尖锐的形体来源于犹太人的传统符号——"大卫之星"（Star of David），这颗断裂的星如被击碎一般曲折地散落在场地中，隐喻着犹太人经受的磨难。在叙述这段历史的过程中，建筑的规则感和稳定感总是被其形体上的切槽和"伤口"所破坏，光线与结构部件突兀地切入内部空间，划破建筑的表皮，有时候甚至直接打断了形体。建筑中设置了三条叙事走廊（narrative promenades）——分别代表着大屠杀时期德国犹太人不同的生活与命运[1]——有两条长廊故意被设置成了逼仄的死胡同，令人感到压抑与冰冷。里伯斯金试图通过这栋建筑表现出的希望、焦虑与缺失感来表达德国犹太人的复杂历史，沉浸式的场所体验直抵人心，远胜于文字资料的说教。

1989-1999

东经 13°23′44″

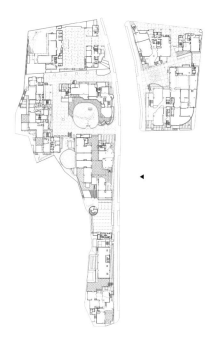

平面图

立面图

69

HILLSIDE TERRACE COMPLEX I-VI
FUMIHIKO MAKI (MAKI & ASSOCIATES)
Tokyo, Japan

代官山集合住宅
槙文彦（槙综合计画事务所）
日本，东京
1967—1992 年

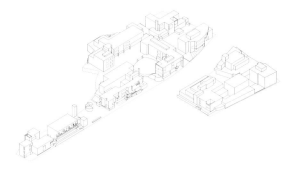

代官山集合住宅项目起始于 1967 年，在近 30 年的设计周期内一共经历了 6 个阶段的发展，不断地呼应着城市的变化。它集合了住宅、商业和文化设施，是一处多功能、中等密度的城市开发项目，槙文彦在此提出了一系列与公共生活密切相关的实验性设计策略。随着时间的推移，东京的城市文化与设计者的观念都发生了转变，采用的建筑处理手法也不尽相同，但整个项目的设计意图仍然保持着连贯性，例如公共功能始终都安排在一层，而生活空间与私人空间则一直安排在建筑上部。集合住宅中的广场、通道、植被与开放空间和建筑本体一样，都经过了建筑师精心的策划与安排。

1967-1992

150 /

北纬 41°03′16″

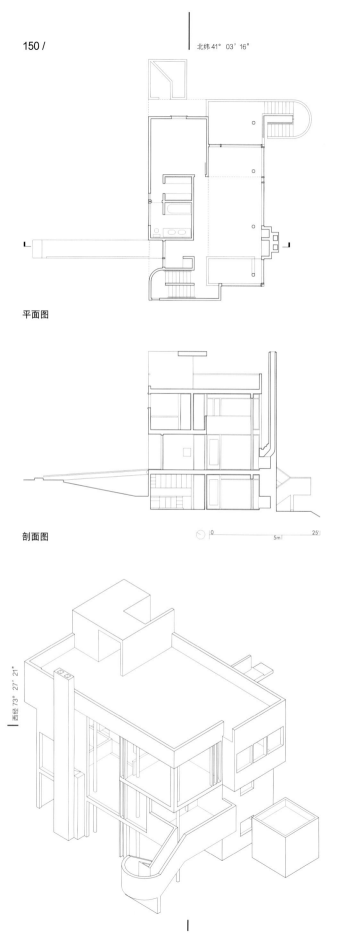

平面图

剖面图

西经 73°27′21″

SMITH HOUSE
RICHARD MEIER
Darien, Connecticut, USA

70

史密斯住宅
理查德·迈耶
美国，康涅狄格州，达里恩
1965—1967 年

　　史密斯住宅体现了抽象与理想主义的设计观念，它以纯白的外观及清晰的几何构成关系与周围青翠的海岸线风景形成了鲜明对比。入口前的步道将人引向一面白色的高墙，较为私密的空间都依附在这面封闭墙体旁边；走入白色高墙中部的开口便进到了建筑内部，迎面可见一片连续而宽阔的玻璃幕墙，周围树林和水面映入眼帘，一览无余。两条分置的走道组织并串联起了水平交通流线，垂直交通则由两部楼梯承担——一部位于建筑内的拐角位置，另一部则突出于形体，二者的对角线构图关系为建筑平面增添了一股张力。总体而言，史密斯住宅可以看作勒·柯布西耶的纯粹主义（Purist）建筑实践的延伸，同时也顺应了美国的地形环境、材料质感与家庭生活方式。从外部望去，史密斯住宅具有雕塑般的观感；身处其内，建筑则变成了欣赏周围风景的平台。

1965-1967

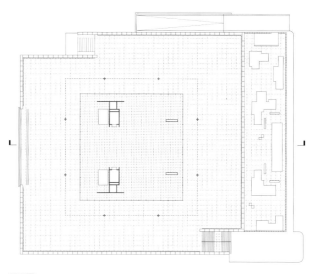

平面图

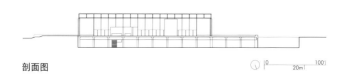

剖面图

71

NEW NATIONAL GALLERY
MIES VAN DER ROHE
Berlin, Germany

德国国家美术馆新馆
密斯·凡·德·罗
德国，柏林
1962—1968 年

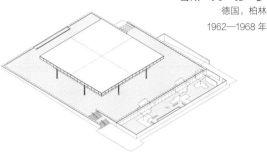

  1961 年，密斯受邀设计了德国国家美术馆新馆。游子还乡，这栋建筑是密斯献给阔别多年祖国的赠礼，也是密斯本人的收山之作。国家美术馆新馆拥有完美的布局和比例，是密斯"放置在台基上的玻璃盒子"这一设计概念的终极版本。在那个时代，大多数人还将博物馆视作"封闭的艺术堡垒"，而密斯却设计出一个"全玻璃的灯塔"——它如同一片完全透明的观景平台，嵌在了白色的花岗石基座和沉静的黑色钢结构屋顶之间。建筑周围的场地空旷而开敞，密斯在此精心地设计了景观视野：人们登上美术馆台基后向北瞭望，便能看到奥古斯特·施蒂勒（August Stüler）设计的圣马修教堂（St. Matthew's Church）及汉斯·夏隆设计的柏林爱乐音乐厅。走进美术馆，通过楼梯向下行进便可进入基座的内部，其中布置了各项服务设施和较为传统的画廊。尽管后来场地周围建造了不少浮华的晚期现代主义建筑，国家美术馆新馆却依旧静谧地矗立在台基之上，不为所动。

1962-1968

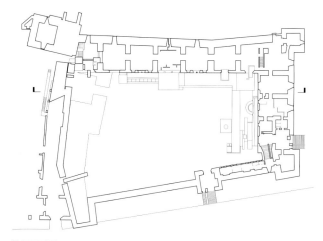

首层平面图

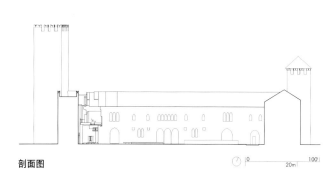

剖面图

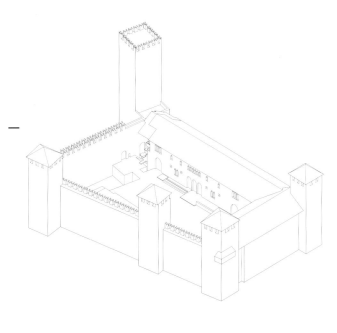

北纬 45°26'23"

CASTELVECCHIO MUSEUM
CARLO SCARPA
Verona, Italy

72

古堡博物馆
卡洛·斯卡帕
意大利，维罗纳
1956—1964 年

斯卡帕将维罗纳的一座古堡垒修复整理后，改建为一座市民文化建筑——这便是古堡博物馆。建筑师在复原历史建筑布局与引入现代建筑语汇之间发起了一场审慎的对话，传统材料与现代材料并举，大理石、钢材、黄铜与木材显得极为契合。斯卡帕决定拆除北部营房的最后一间及室外的大楼梯，引发了一定争议[1]，但他的处理方式却令人叹服：拆除楼体后，空出一处通高空间，在下方地面铺设了混凝土平台，并用一部"L"形的混凝土支座将中世纪的骑马者雕像高高举起，在空间上方修建了一跨人行天桥，以此形成了一种立体的视觉空间维度。古堡博物馆中遍布着精妙绝伦的细部设计，例如画廊入口处分离的地板与墙壁，以及对大型展品附近围栏的消解处理。路易斯·康曾赞美道："细部倾诉着对本质的赞叹。（The detail is the adoration of Nature）"

1956-1964

东经 10°59'16"

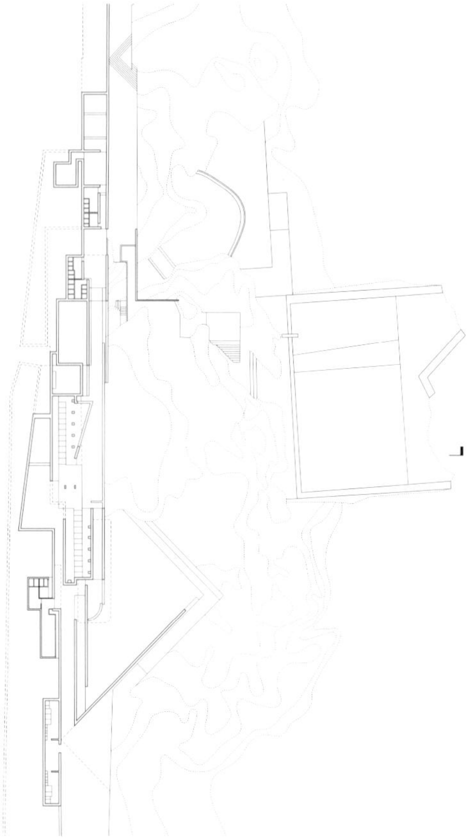

平面图

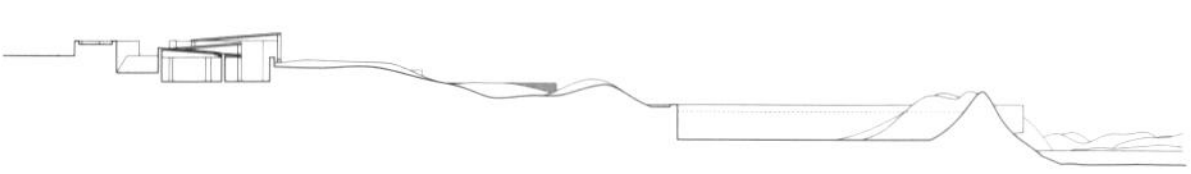

剖面图

西经 8°42′27″

北纬 41° 11' 34"

PISCINA DE MARÉS
ALVARO SIZA
Palmeira, Portugal

73

莱萨浴场
阿尔瓦罗·西扎
葡萄牙,帕尔梅拉
1961—1966 年

　　从马托西纽什(Matosinhos)繁忙的滨海公路旁放眼望去,你几乎注意不到莱萨浴场的存在。长长的海堤旁退让出一处平缓的斜坡,由于透视角度的误导,斜坡显得不是很深;沿着斜坡向下漫步,首先看到了广阔的大海,建筑屋顶的轮廓也逐渐显露了出来,然后就进入了内部的更衣室;继续前行,穿过浸脚池后便可看到设置于海岸礁石间的宽阔泳池,混凝土、木材与崎岖的礁岩相互拼插并融为一体。走下建筑的水泥台阶,踏上松软的沙滩,两座游泳池浮在地上,似乎随时会没于海水之中。西扎的设计方案既大胆又精妙,既简洁又复杂,处处流露出令人愉悦的建筑体验。

1961-1966

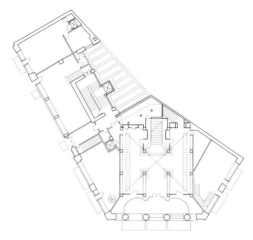

首层平面图

立面图

1909-1911

MICHAELERPLATZ HOUSE
LOOSHAUS
ADOLF LOOS
Vienna, Austria

74

米歇尔广场公寓（路斯公寓）
阿道夫·路斯
奥地利，维也纳
1909—1911 年

路斯公寓虽然处在巴洛克风格盛行的维也纳历史中心区，其立面却表现出明显的简化特征：它将材料本身的特性——纹理、光泽、色调与质地作为建筑唯一的装饰加以表现，这样的手法在当时引起了一定争议。这栋建筑是为奢侈服装品牌 Goldman & Salatsch 设计的总部楼（路斯一直致力于该品牌的推广），楼下是用于营业的商铺，上层则是公司的办公室和公寓。这栋建筑清晰地表现出路斯对于时尚与建筑形式的观念，建筑入口处设置了装饰性的柱体，大理石的材质纹理与白色的墙面抹灰本身便构成了一种对比；充满异域风格的抛光木材几乎没有做任何造型处理，房屋内则由简单的玻璃橱柜划分出空间与功能。最重要的一点是，路斯公寓塑造出一种全新的城市建筑形象，以其新颖的形式将充斥着菲舍尔·冯·埃拉赫[1]巴洛克式样的米歇尔广场变成了一处商业与公共生活的现代空间。

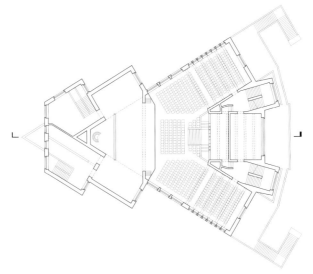

平面图

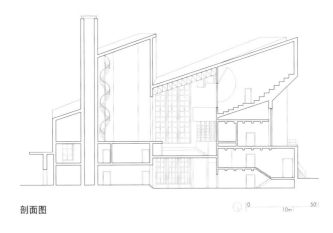

剖面图

1927-1929

RUSAKOV WORKERS' CLUB
KONSTANTIN MELNIKOV
Moscow, Russia

75

鲁萨科夫工人俱乐部
康斯坦丁·梅尔尼科夫

俄罗斯,莫斯科
1927—1929 年

　　梅尔尼科夫设计的工人俱乐部是短命却又影响深远的构成主义(constructivism)为数不多的现存作品之一,这栋建筑试图表现群众的革命力量与坚强韧性,其体块呈现出肌肉紧绷的力感,同时强调出向上的动势。位居建筑中心的是中央舞台,钢筋混凝土浇筑的观众席被分成三部分,以大胆夸张的扇形布局从主体中飞跃而出——这些引人注目的楔形体块像是巨型的齿轮,实际上是观众席倾斜地面的忠实表现。舞台以及后部空间采用了砖材,从背面看起来更像是一座典型的苏联建筑,这令俱乐部主要的使用者们——电车工人倍感亲切。在建筑的一层走廊中布置着移动的隔板,为建筑功能带来了一定灵活性。

162 /

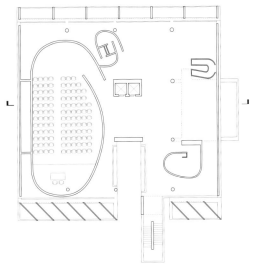

平面图

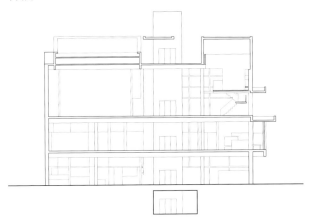

剖面图

MILL OWNERS' ASSOCIATION BUILDING
LE CORBUSIER
Ahmedabad, India

76

棉纺织协会总部
勒·柯布西耶
印度,艾哈迈达巴德
1951—1954 年

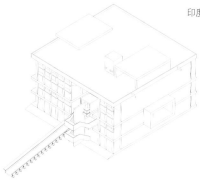

在棉纺织协会总部的设计中,勒·柯布西耶的"新建筑五要点"取得了关键性的进展:这栋建筑不但考虑了半干旱地区独特的地方特色,还适应了当地的季风性气候条件。沿着一条笔直的坡道漫步而上,便来到了开敞的门厅空间;顺着宽敞的楼梯向上层行走就进入了集会大厅,自由的平面被充满雕塑感的建筑构件所装饰。入口立面朝西,柯布西耶为了防止西晒的影响,在房间外布置了一套独立的"遮阳板",并根据当地日照情况调整了混凝土板的倾角:冬季温暖的阳光可以射入建筑内部,而夏天混凝土板可以遮蔽强烈的光线。建筑的东立面由轻质框架构成,从附近河面上吹来的凉风可以直接流入室内;屋顶上有一扇"U"形的混凝土天窗用于采光,还设置了若干水池和屋面种植措施以调节室内温度。

1951-1954

北纬 34°06'02"

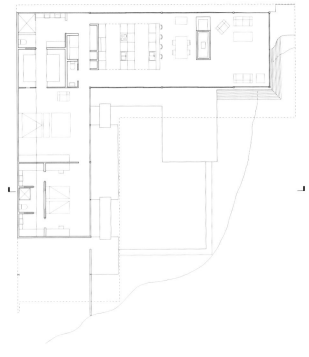

平面图

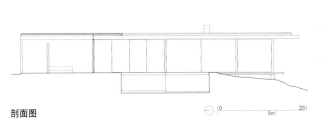

剖面图

西经 118°22'13"

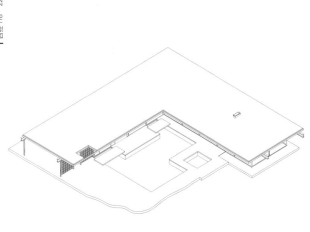

77

STAHL HOUSE
CASE STUDY HOUSE NO.22
PIERRE KOENIG
Hollywood, California, USA

斯塔尔住宅（第 22 号案例住宅）
皮埃尔·凯尼格
美国，加利福尼亚州，好莱坞
1960 年

在约翰·恩腾扎（John Entenza）发起的"案例住宅计划"[1]的一系列作品之中，斯塔尔住宅无疑是最上镜的。自从朱利叶斯·舒尔曼拍摄了那张广为流传的照片，[2]斯塔尔住宅便成为 20 世纪洛杉矶现代生活场景的代表。这座简洁的玻璃住宅坐落在好莱坞山（Hollywood Hills）一块隆起的地块上，顶部使用了极轻薄的预制钢板作为屋面，它不仅限定出住宅空间的高度，深远的挑檐也为连续不断的玻璃幕墙遮蔽了阳光；"L"形的平面布局中表现出建筑与泳池、天空和景观之间的相互关系，卧室被布置在一面不透明的钢结构墙体一侧，与一旁的公路隔绝开来；开放而灵活的起居空间则被悬挑在基地边缘，透亮的玻璃墙让整个生活空间戏剧性地融进了洛杉矶的天际线。

— 1960

北纬 40°46'25"

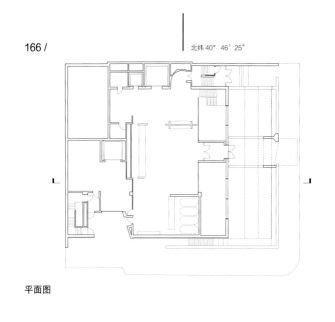

平面图

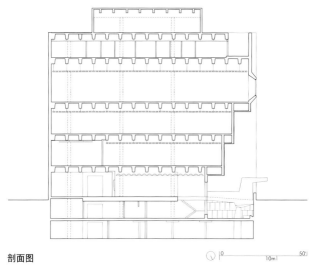

剖面图　　　　　　　　　　0　10m　50

西经 73°57'50"

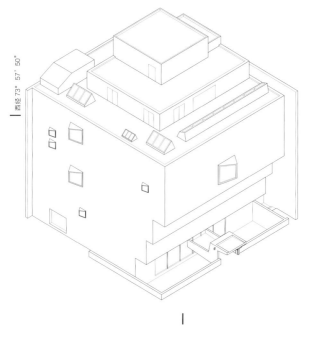

THE BREUER BUILDING
WHITNEY MUSEUM OF AMERICAN ART
MARCEL BREUER and HAMILTON P. SMITH
New York, New York, USA

78

布罗伊尔楼（惠特尼美国艺术博物馆）
马塞尔·布罗伊尔与汉密尔顿·P·史密斯
美国，纽约州，纽约
1963—1966 年

    马塞尔·布罗伊尔设计的惠特尼美国艺术博物馆，其形体非常大胆，包覆着花岗石的建筑体量向着街道逐层外挑，窗户也从立面中斜突出来，表现出紧致的力感。这栋建筑是惠特尼博物馆的第三处展馆，封闭的建筑表皮内包覆着一系列复杂的开放空间，等待着公众前来探访。惠特尼美国艺术博物馆犹如一块巨大的独石，力贯千钧地砸入了陨石坑一般的下沉广场之中，其规模、手法、意图与形象都与遍布纽约的玻璃塔楼相去甚远。倒立的金字塔形体为建筑底层拓展出更大的室外空地，提升了底层空间的品质，逐层增进的上部形体中则包含着巨大的展厅。建筑师尽力拓展展厅面积并为游客提供了开放的观展空间，看似倾斜的窗户是在迎合特意设置的景观视角，一部优雅的楼梯和超大的电梯将各个空间串连在一起，引导着游客在建筑内探索。2014 年，惠特尼美国艺术博物馆搬迁至由伦佐·皮亚诺设计的新馆内；2016 年，布洛伊尔楼变成了纽约大都会艺术博物馆的卫星馆，此后便被称作"大都会艺术博物馆布洛伊尔楼"（The Met Breuer）。

1963-1966

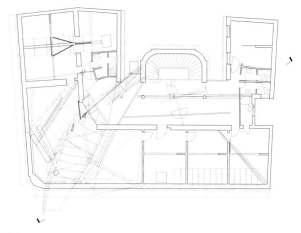

平面图

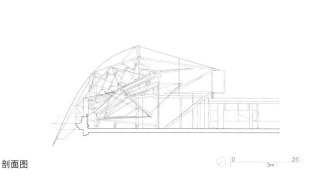

剖面图

ROOFTOP REMODELING FALKESTRASSE
COOP HIMMELB(L)AU
Vienna, Austria

79

屋顶加建
蓝天组
奥地利，维也纳
1983—1988 年

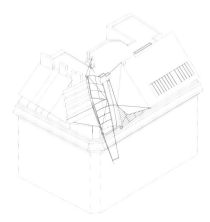

　　在维也纳历史中心区域的一栋毕德麦雅时期¹建造的历史建筑上，蓝天组设计了一项极为大胆的屋顶加建项目，这项作品不但让蓝天组声名鹊起，也激励了世界各地的建筑师挑战建筑设计的"标准程序"。按照设计者的构思，屋顶加建如同一束"由街道直接贯穿建筑屋面的能量射线"（毫无疑问，这项构思受到了建筑旁边"猎鹰街"的名称影响），一条紧绷的钢索与原有建筑的女儿墙相交，划出一道漂亮的弧线，它穿透并扩展了屋顶平面，将不同形状的构件连为一体。开敞、平滑的表皮与封闭、折叠的表面交织在一起，营造出复杂的空间感受和光影效果，难怪建筑师们将屋顶加建称作"桥梁与飞机的混合体"。

1983-1988

北纬 42°52'38"

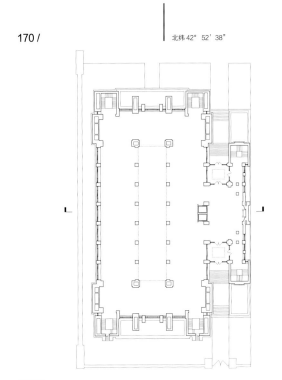

平面图

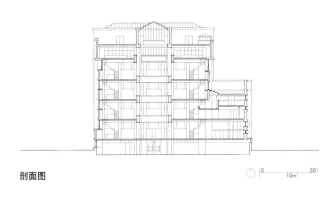

剖面图

西经 78°51'08"

—— 1902-1906

LARKIN COMPANY ADMINISTRATION BUILDING
FRANK LLOYD WRIGHT
Buffalo, New York, USA

拉金公司行政大楼
弗兰克·劳埃德·赖特
美国，纽约州，布法罗
1902—1906 年，1950 年拆毁

  拉金公司行政大楼展示了建筑师赖特与业主拉金的一条共同信念：劳动生产是社会契约的基础。从外观看，建筑立面中雄壮的支墩携着一股古埃及神庙的宏伟气势，内部空间则展现出赖特作品的典型风格。支墩与墙壁、空间与体量，收放之间皆体现了有机的平衡关系。为了保证形式的统一，赖特甚至使用红砖与红色砂浆将建筑的钢结构包裹起来。通过迂回的入口空间便来到建筑中央五层高的"工作空间中庭"（central work-space atrium），光线穿过顶部的玻璃天花板照亮了由奶油色砖块组成的支柱与壁龛，整个空间犹如一座灿烂的工人大教堂，熠熠生辉。拉金公司行政大楼中采取了一系列新创的建筑设施，如中央空调、内置桌面的办公家具以及悬箱式水冲厕所，服务空间和消防楼梯被赖特安排在支墩内部，这在当时也是很新颖的处理手法。

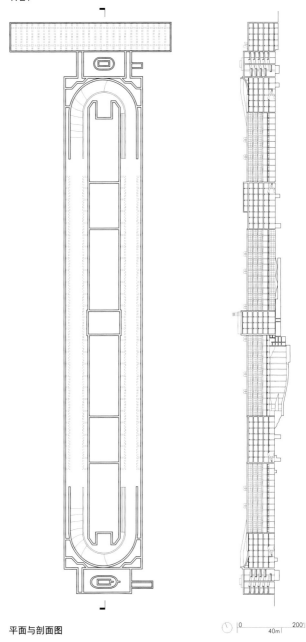

平面与剖面图

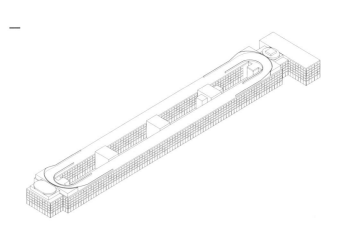

FIAT WORKS
GIACOMO MATTÉ-TRUCCO
Turin, Italy

81

菲亚特工厂
贾科莫·马泰-特鲁科
意大利，都林
1915—1921 年

菲亚特工厂是一栋前所未见的革命式建筑。它的屋顶上居然设置了一圈汽车跑道，车辆通过巨大的混凝土螺旋坡道就可以开上屋顶测试性能！漫长的形体之内隐藏着四处巨大的内庭，外立面上的钢筋混凝土框架也被毫不遮掩地表现出来。可以说，这项作品为 20 世纪早期的建筑师提供了一种鲜明的形式参照，鼓舞了他们的雄心壮志。各种生产材料从建筑底层进入车间，随着汽车制造过程的推移逐渐到达建筑上层空间，组装完成之后便可开上屋顶的跑道；作为最早采用的垂直化生产方式的城市工厂，这栋建筑率先采用了大跨度钢筋混凝土结构，并将汽车的转弯半径尺度与高效的劳动力组织结合起来。菲亚特工厂出身于未来主义者追求速度的狂想，在勒·柯布西耶的《走向新建筑》(*Toward an Architecture*)一书中获得了声望。它是集中式生产与劳动效率的纪念碑，同时也代表着坚韧的功能主义精神与浪漫的工业化理想。

174 /

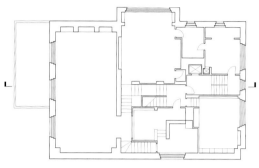

平面图

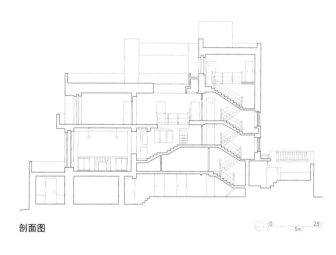

剖面图

1928—1930

VILLA MÜLLER
ADOLF LOOS
Prague, Czech Republic

82

穆勒别墅
阿道夫·路斯
捷克，布拉格
1928—1930 年

  穆勒别墅在早期现代主义的形式论战中占据了核心位置。这项作品集中体现了路斯的建筑理论，不但表现出其对于装饰的观念，也展示了他的空间操作方法。在路斯的观念中，一栋建筑的外部应当是简洁而沉静的，内部则应当表现出丰富的材质特性与空间品质。穆勒别墅的四个立面无视并嘲弄了"正统的对称形式"，并以不对称的开窗与阶梯式的体量搅动着"纯粹的对称法则"。一旦进入建筑致密而压缩的入口空间，便可看到两部位置不对称的楼梯各自向上攀升——这两条交通流线体现出截然不同的组织策略，使公共空间与私密空间、大空间与小空间之间发生了一系列连续的转化，辅助空间则依次排布在主要空间的周围。穆勒别墅中的房间形状很明确，同时还具有丰富的材质与鲜明的色彩，它们在不同的楼层中拼接契合，彼此之间依靠蜿蜒曲折的楼梯发生联系。

176 /

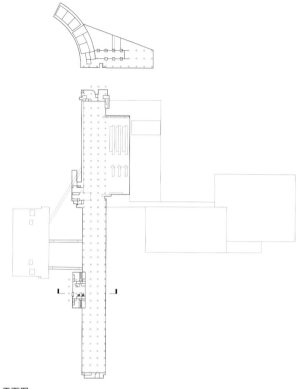

平面图

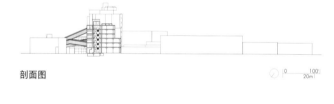

剖面图

1926-1930

VAN NELLE FACTORY
JOHANNES BRINKMAN, LEENDERT VAN DER VLUGT, and MART STAM
Rotterdam, The Netherlands

范内勒工厂
约翰内斯·布林克曼、伦德特·范德与马尔特·斯塔姆

荷兰，鹿特丹

1926—1930 年

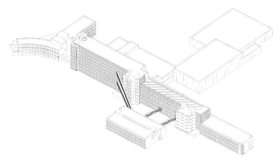

如果说效率、性能和功能主义定义了早期现代主义的政治使命，阳光、空气和工人福利定义了荷兰现代主义者的社会使命，那么范内勒工厂就是这两种精神混合之后的典型代表。这栋工厂是为了加工咖啡、茶叶和鼻烟等细颗粒产品而建造的，为了控制生产过程中的交叉污染，严格限制了生产公差并设计了一套封闭的循环系统。在厂房内部，整个生产过程被清楚地展示出来：原材料从建筑顶层送入厂房，经过逐层加工后再经由暴露在外的传送管道分配到包装车间。建筑的混凝土框架被一大片玻璃幕墙包裹起来，这在当时着实算是一件创举了。透明的玻璃墙不但展示了生产过程，同时也为工人们提供了安全舒适、干净清洁、采光良好的工作环境。范内勒工厂既被功能主义者所欣赏，也受到了构成主义者的喜爱，实属罕见。

178 /

北纬 42° 18' 13"

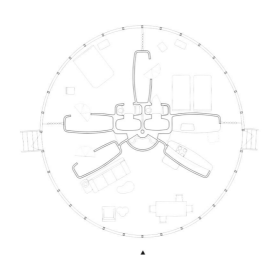

首层平面图

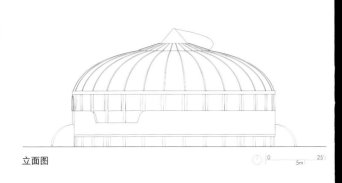

立面图

西经 83° 14' 03"

1927

DYMAXION HOUSE
BUCKMINSTER FULLER
Dearborn, Michigan, USA

84

动态高效住宅
巴克敏斯特·富勒
美国，密歇根州，迪尔伯恩
1927年设计，1945年改进并建造

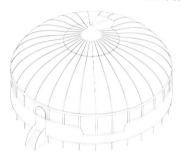

　　巴克敏斯特·富勒将"dynamic"（动态）、"maximum"（最大化）与"tension"（张力）三个词汇组合起来，创造出一个全新的词语："Dymaxion"，也就是动态高效住宅。这项发明旨在将汽车生产线、工业材料与融资模式引入家庭住宅产业中。动态高效住宅是一种极富远见的可持续建筑，也是未来的"生活机器"，它能够自给自足，维护低廉成本，并且可以大规模预制生产。动态高效住宅采用工程材料建造（主要是铝材），不但能抵御地震和飓风的侵袭，还无需定期喷涂翻新。圆形的建筑平面围绕在核心部件周围（其中包括了主结构支柱和公用生活设施），将建筑的占地面积缩减到最小；可移动的"服务舱"灵活多变，能够根据不同的使用需求调整空间布局；圆柱状的建筑形体减少了热能损失，同时也节约了材料——整栋房屋只有1.5吨重，仅仅是普通住宅的百分之一而已。

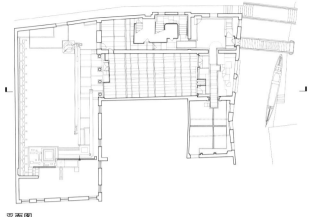

平面图

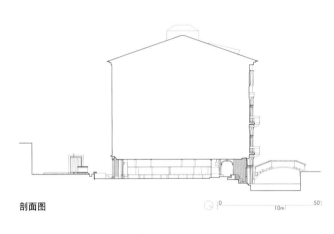

剖面图

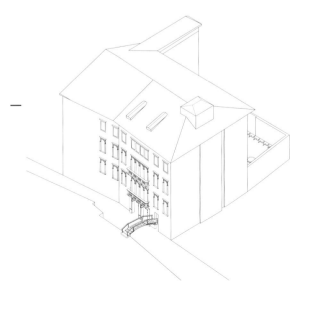

QUERINI STAMPALIA RENOVATION
CARLO SCARPA and CARLO MASCHIETTO
Venice, Italy

奎里尼·斯坦帕利亚基金会更新
卡洛·斯卡帕与卡洛·马斯基耶托

意大利，威尼斯

1961—1963 年

斯坦帕利亚基金会更新项目位于威尼斯一栋历史悠久的 16 世纪府邸的底层。与古堡博物馆一样，此次设计也需要处理一系列历史场景问题，不同的是这一次斯卡帕还需要面对"水的问题"——建筑内外都与水发生着密切的关联。这座古宅位于威尼斯大运河旁边，长期受周期性洪水的影响[1]，因此更新方案在设计之初就必须承受并应对洪水的侵扰。一座钢木结构的拱桥横跨在运河之上，将建筑入口与外部广场联系起来，几间新设的公共空间嵌套在建筑古老的结构之中，虽然身处其内，平面与走道并没有完全受旧墙的约束。同时，斯卡帕在建筑内还使用了其标志性的"三色马赛克"装点了地面与墙体，塑造出独特的艺术效果。[2] 经过前厅和主展厅，便来到建筑后方的一座小花园，狭窄的渠道引着一条涓涓细流穿过整个庭院，水流灌入了枯井旁的一方圆形小池后便了无踪迹。

1961-1963

北纬 41° 50′ 01″

平面图

剖面图

西经 73° 19′ 21″

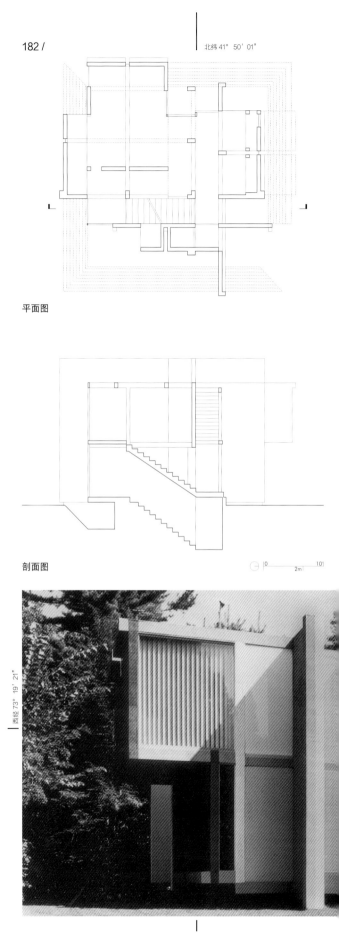

86

HOUSE VI
PETER EISENMAN
Cornwall, Connecticut, USA

六号住宅
彼得·埃森曼
美国,康涅狄格州,康沃尔
1972—1975 年

彼得·埃森曼设计的六号住宅是一系列图解的结果。这栋住宅通过对于建筑"自洽性"(autonomous)可能的探索,推动了第二波后现代建筑浪潮。[1] 所谓"建筑的自洽性",是指一套基于建筑学学科范畴的任意(arbitrary)而独特(unique)的空间语言(language),同时还将建筑整体作为一套独立的辩证关系加以理解并作出诠释。六号住宅几乎与场地毫无关联,这却更加凸显出其本身内涵的"自洽",建筑中的颜色与网格、实体与虚空都是基于空间操作(operations)的方法性陈述,四面墙体在碰撞拼接中形成一种垂直交错的复杂关系,建筑的空间与形体便从中诞生了。埃森曼破坏并"陌生化"(Defamiliarize)了家庭环境中的一些基本功能,以此挑衅常规的空间认识:柱子将餐桌一分为二,玻璃槽则把卧室里的大床劈成两半;空隙空间只能远观却无法进入,甚至连一部楼梯都是反转颠倒的。

1972-1975

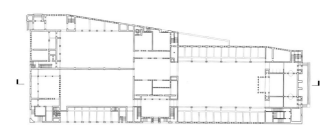
平面图

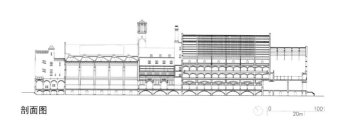
剖面图

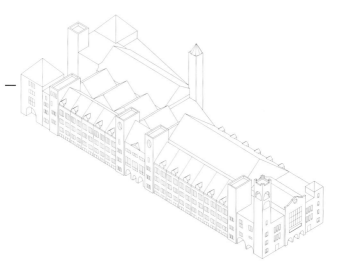
1896-1903

BEURS VAN BERLAGE
HENDRIK PETRUS BERLAGE
Amsterdam, The Netherlands

87

贝尔拉赫证券交易所

亨德里克·彼得鲁斯·贝尔拉赫

荷兰,阿姆斯特丹

1896—1903 年

19 世纪的集仿建筑美学令贝尔拉赫感到不适,他尤其厌恶荷兰本土矫饰的折中主义建筑。贝尔拉赫证券交易所是这座城市中规模最大的建筑之一,设计师在设计中采用了"新罗曼式风格"(neo-Romanesque style),他感到这所建筑必须要传达出一种"安全感",同时也必须表现出自由和开放的特性。证券交易所的形体组织非常清晰,展示出贝尔拉赫根深蒂固的"多元统一"(unity in diversity)的世界观。在建筑本体与其内部三座大厅的设计中,普通砖块与釉面砖、粗凿的毛石与抛光石板形成材质上的对比,砖石结构的支墩也与表现着维奥莱·勒杜克(Viollet-le-Duc)理性主义技术的裸露铁质桁架极为契合。证券交易所既保持了简单合理的建筑规划,同时也避免了过度的建筑装饰,贝尔拉赫通过这项作品将荷兰建筑从资本驱动的折中风格引向了现实主义风格(Dutch Realism),并最终引发了荷兰现代主义建筑的产生。

186 /

北纬 34°09'07"

首层平面图

立面图

0　5m　25'

东经 118°09'39"

1907-1909

GAMBLE HOUSE
GREENE & GREENE
Pasadena, California, USA

88

甘布尔住宅
格林兄弟

美国，加利福尼亚州，帕萨迪纳

1907—1909 年

　　格林兄弟设计的甘布尔住宅是甘布尔家族（Gamble family，日用消费品巨头宝洁公司的联合创始人）的一所疗养住宅，也是美国艺术与手工艺运动（Arts and Crafts movement）的代表性作品。这栋建筑体现了"总体艺术"[1] 的概念——房子里面的每个元素都是由格林兄弟定制设计的，姑且不论大楼梯上精致的木雕花纹和蒂芙尼（Tiffany）专门制作的彩色玻璃窗，甚至连灯笼、开关面板和家具全都是量身打造的。甘布尔住宅体现了加利福尼亚建筑的开放特征，还融合了日本传统建筑的风格（尤其体现在细木工的设计细节上）。房屋的内部与外部空间联系紧密，还结合景观设置了小憩的门廊与若干阳台，使主人可以充分享受加州健康怡人的气候条件。

188 /

北纬 33°36′21″

平面图

剖面图

1922-1926

西经 117°55′04″

LOVELL BEACH HOUSE
RUDOLF SCHINDLER
Newport Beach, California, USA

89

洛弗尔海滨住宅
鲁道夫·申德勒
美国，加利福尼亚，纽波特比奇
1922—1926 年

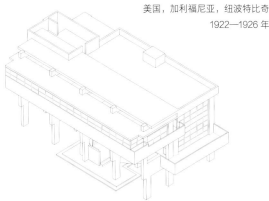

在宣言式的自宅建成四年之后，申德勒又设计了洛弗尔海滨住宅。该方案延续了建筑师对于结构、生活空间与交通流线的探索，同时，申德勒还尝试使用当时很罕见的现浇混凝土来建造这栋房屋。住宅的主人菲利普·洛弗尔博士（Dr. Philip Lovell）提倡通过健康的居住环境实现清洁的生活方式，广受美国媒体的推崇 [（之后他又委托里夏德·诺伊特拉 (Richard Neutra) 设计了洛弗尔健康之家）]。申德勒以五跨平行的现浇框架作为住宅空间与功能的骨骼，将主要体量举到空中，并用由此形成的门廊调和了公共与私密空间。界限明确的框架内部布置着住宅的生活区域，两层通高的起居空间极为开敞，混凝土结构被木材包裹后充满生活气息，主人的私密空间则被高高举起，安置在建筑的顶层。

190 /

平面图

剖面图

北纬 45°02′58″

TURIN EXHIBITION HALL
PIER LUIGI NERVI
Turin, Italy

90

都林展览馆
皮尔·路易吉·内尔维
意大利，都林
1948—1954 年

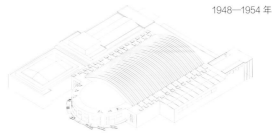

在都林展览馆的设计中，内尔维以浅拱结构（shallow arch configuration）建造出宽达 94 米的无支柱模块化钢筋网混凝土拱顶，这座宏大的屋盖是精湛的技术能力与艺术表现力的结合，可谓"技艺合一"，甚至还可以透过结构之间的空隙达到顶部采光的效果。现场浇筑的拱基托举着巨大的顶板，顶板向上又延伸分裂为三段弧形的拱板，它们由一系列预制波浪形钢筋混凝土元件组成，每一块的尺寸约为 5 米 ×2.5 米，厚约 5 厘米。预制拱板上，沿对角线位置设置出窗洞，日光可以由此射入内部展厅。为了增强屋顶水平方向的结构刚性，内尔维还在拱板间加入了"V"形的三角加劲肋，同时使用现浇混凝土将每个波浪拱板的顶点与卡槽固定起来，各个构件被锁定成一个连续的整体。巨大的弧形拱基不但支撑着顶板，还承托着展厅二层的平台。展厅尽端设置了一处直径达 40 米的半圆形大厅，其上的半球穹顶由钢筋混凝土与菱形拱肋浇筑，这座宏大的穹窿是整个展厅空间序列的终结。

东经 7°40′53″

1948-1954

平面图

立面图

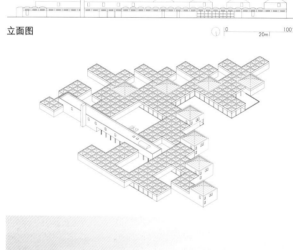

MUNICIPAL ORPHANAGE, AMSTERDAM
ALDO VAN EYCK
Amsterdam, The Netherlands

阿姆斯特丹市孤儿院
阿尔多·范艾克

德国，柏林

1955—1960 年

阿尔多·范艾克是"第十次小组"[1]的发起人之一，他在阿姆斯特丹郊外的一处孤儿院项目中找到了完美的实践机会：这栋单一而连续的"毯式建筑"[2]反映出明显的社群（community）属性，并使用"市场"、"街道"与"广场"的空间意象激发出人文主义内涵。该建筑大部分由砖材砌筑，屋顶采用 3.3 米 ×3.3 米（小房间）与 10 米 ×10 米（活动室）规格的方形预制混凝土薄壳模块，所有房间与外部庭院空间的尺寸也都是遵照屋顶的模度确立的。阿姆斯特丹市孤儿院分为八翼，各自布置了生活空间与教室，应对了不同年龄与性别儿童的需求。虽然建筑的整体布局如同机器一般工整，但建筑内的各个部件——如长凳、门廊、拱道与窗台，却都是遵照儿童的身体尺度精确设计，连建筑内的镜子与窗户都被安排在令人意想不到的位置——只有小孩子才能找到它们。一般而言，社会福利类建筑（孤儿院、养老院）都因更加注重使用效率而忽略了建筑的活力，范艾克却设计了许多开放空间，甚至还预留出一部分暂时无用的场地。在此处，建筑师不仅建造了一所孤儿院，更是在创造一座"微缩的村落"。

1955-1960

北纬 42°22′25″

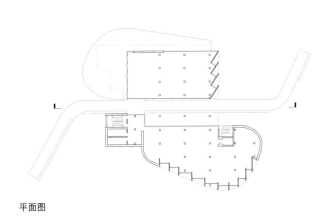

平面图

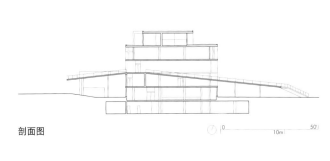

剖面图

西经 71°06′52″

CARPENTER CENTER FOR THE VISUAL ARTS
LE CORBUSIER
Harvard University, Cambridge, Massachusetts, USA

92

卡朋特视觉艺术中心
勒·柯布西耶
美国，马萨诸塞州，剑桥
1959—1962 年

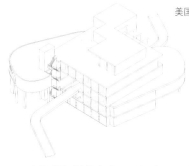

  卡朋特视觉艺术中心是柯布西耶在美国所做的第一栋作品，它是一部"建筑化"的《勒·柯布西耶全集》[1]，将大师的设计手法淋漓尽致地展现出来。视觉艺术中心坐落在哈佛大学旁边的街道一角，呈立方形体量，整体布局大致与周围的街道保持平行；柯布西耶将外挂着混凝土遮阳板的建筑主体扭转了一定角度，以此调和建筑与旁边的博物馆和小型教师俱乐部的空间体量关系。一条漫长而蜿蜒的"S"形坡道贯穿了整片场地，它逐渐抬升并将建筑从中心位置一分为二。这条步道吸引着校园中的行人走向建筑，在漫步中发现并参与视觉艺术中心内举办的各项活动。建筑师还依着建筑体量和流线的驱动，在现浇混凝土的主楼旁边布置了两组外突的弧形体块，并在架空的体块下方设置出艺术中心的次入口和室外展场，为到访艺术家准备的临时公寓则被安置在屋顶平台上，不会受到下部公共空间的侵扰。

1959-1962

平面图

剖面图

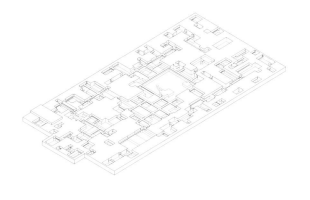

93

FREE UNIVERSITY OF BERLIN
CANDILIS-JOSIC-WOODS and MANFRED SCHIEDHELM
Berlin, Germany

柏林自由大学

康迪利斯-若西克-伍兹与曼弗雷德·席德赫姆

德国,柏林

1963—1973 年

　　柏林自由大学是由"第十次小组"成员所建造的第一栋建筑规划项目,也是目前为止被引用最多的"毯式建筑"。这项作品以独特的方式实践了谢德拉克·伍兹[1]的"干茎"(stem)与"网络"(web)理论,建筑师基于一套水平的网格体系构架了建筑的平面并塑造出一个"微缩的密集城市",意图复制北非民居(casbah)中的"分阶化"(scalar)与"经验化"(experiential)的多样空间体验,并以此呼应了阿尔多·范艾克在校园规划设计中倡导的"迷宫似的清晰"(labyrinthine clarity)的设计概念。建筑大多只有一层高,平铺在看似无穷无尽的网格之中,若干位置点缀着一些二三层高的楼房,其间穿插的层层密密的花园与内庭为室内带来了明亮的光线。柏林自由大学是一座"没有层级的矩阵建筑"(nonheirarchical matrix building),以模块化的结构与围护层所带来的设计弹性为基础,通过便于重新配置的分区与结构[包括柱、楼板与简·普鲁韦(Jean Prouvé)设计制造的科尔坦金属围护墙]提供出可变与灵活的丰富空间。

1963-1973

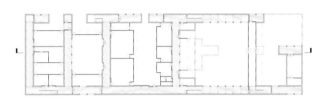

平面图

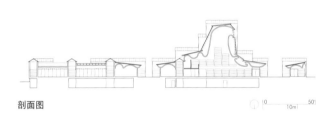

剖面图

北纬 55° 45′ 42″

BAGSVÆRD CHURCH
JØRN UTZON
Copenhagen, Denmark

94

鲍斯韦教堂
约翰·伍重
丹麦，哥本哈根
1969—1976 年

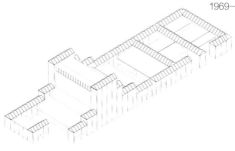

　　鲍斯韦教堂的建筑立面与建造材料看起来与当地的农业建筑或工业建筑非常接近，平面布局则像是丹麦的古城堡或农场庭院，这座宗教建筑完全借由卓越的剖面设计向人们展现出非凡的"神性"。建筑阶梯状的直线造型一目了然，大部分墙体都是白色的预制混凝土板，墙体上部还装饰了白色釉面砖，以此巧妙地提示出内部的剖面空间，屋顶的铝制瓦楞板则进一步展现出一种"工业化的沉静感"。鲍斯韦教堂的剖面中包含着一系列弯曲而翻腾的混凝土现浇"拱壳"，这些曲面拱顶几乎横穿了整个内部空间，在贯入祭坛空间后升腾而起，扶摇而上，直指天穹。北欧平直的太阳光从顶部的天窗射入祭坛，经过拱顶多次漫反射后变成了柔和的光线，盈满了教堂。正如伍重的概念草图所表现的那样，鲍斯韦教堂通过恰如其分的材质表现与巧妙的光线操作，重现了"在白云之下的大自然中举行祭祀"的原初宗教体验。

1969-1976

东经 12° 26′ 41″

200 /

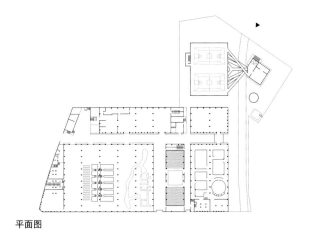

平面图

立面图

西经 46° 41' 00"

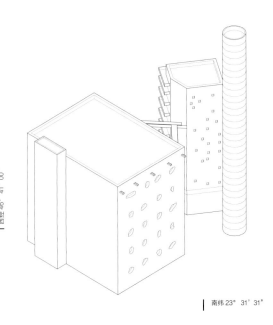

南纬 23° 31' 31"

95

SESC POMPÉIA
LINA BO BARDI
São Paulo, Brazil

庞培娅艺术中心
莉娜·柏巴迪
巴西，圣保罗
1977—1986 年

　　庞培娅艺术中心原本是一座生产金属桶的工厂，改建之后变成了一个庞大的社区生活中心，成了员工的社交、娱乐活动场所。高度较低的厂房被改造成了图书馆、自助餐厅、展览空间和会议室，并且尽可能保持了开放的空间形态；一条原有的雨水渠划分出场地的边界，柏巴迪将两个新建的体育场塔楼布置在水渠两侧，并将漫长的水渠改造成一段"日光浴平台"，塔楼表面形状各异的窗洞被涂成红色，极富特点。新建的现浇混凝土水塔如烟囱一般高高耸立在体育馆旁边，由于不同层面之间采用了"外渗浇筑"（oozing joint）的施工技术[1]，水塔的表皮呈现出不规则的横向条纹肌理，清晰地表现出自身的建造过程。庞培娅艺术中心充分展示出柏巴迪在所有建筑设计中都充分考虑社交互动的建筑概念。

1977-1986

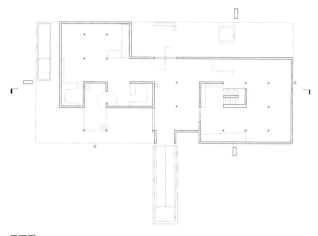

平面图

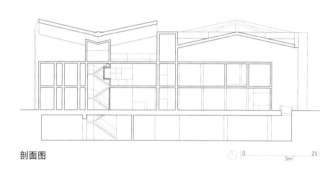

剖面图

CENTRE LE CORBUSIER
HEIDI WEBER HOUSE
LE CORBUSIER
Zürich, Switzerland

96

柯布西耶中心（海蒂·韦伯住宅）

勒·柯布西耶

瑞士，苏黎世

1965—1967 年

海蒂·韦伯住宅是柯布西耶所作的最后几栋作品之一，这栋建筑专门收藏柯布西耶所做的各类艺术作品，集合了公园展馆、画廊和住宅的功能，也是收藏家兼策展人海蒂·韦伯女士的栖身之所。在柯布西耶所有的建成作品中，海蒂·韦伯住宅是唯一一座由内而外全部使用了钢结构的建筑，它采用了模块化的结构设计原则 [其建筑构件遵循柯布西耶的"模度"（Modulor）尺度系统]，各个结构构件全部由构造关系十分清晰的"L"形角钢单元组合而成，角钢经过焊接后便可组成"T"形与十字形的梁柱。建筑立面中的彩釉金属板与玻璃组成的墙壁很有特点，建筑的地面则是包覆了橡胶砖的钢板。海蒂·韦伯住宅中的模块化预制系统其实是柯布西耶早期研究与探索的成果，在其 1939 年所作的"旧金山与列日的法国馆"（the prefabricated pavilion proposed for Liège and San Francisco）、1949 年的"Roq 与 Rob 住宅项目"（the Roq and Rob housing）与 1950 年的"马约门 1950 博览会项目"（the project for an Exposition at the Porte Maillot）中皆可窥见其踪影。

1965-1967

204 /

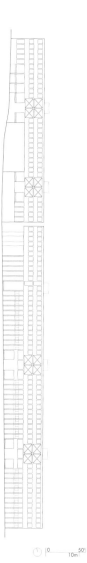

平面图与剖面图

97

GALLARATESE II APARTMENTS
ALDO ROSSI and CARLO AYMONINO
Milan, Italy

加拉拉特西公寓
阿尔多·罗西与卡洛·艾莫尼诺

意大利，米兰
1969—1974 年

罗西一直提倡的，是一种"静默的建筑"，加拉拉特西公寓却发出了振聋发聩的巨响。米兰加拉拉特西的住宅开发项目被委托给建筑师卡洛·艾莫尼诺[1]，慧眼识珠的艾莫尼诺选择罗西作为自己的设计合作伙伴，这是一个极富远见的抉择——罗西设计的 D 栋不仅与艾莫尼诺所作的现代化住宅综合体形成了明确的比照，更如赛罗塔石碑[2]一般，成为对照并理解整个住宅项目的参考线与解读线索，同时，经由 D 栋底层多孔并镂空的公共空间，整个住宅综合体的居民被组织串联在一起。加拉拉特西公寓也许是静默的，但绝非默不作声——这栋建筑清晰地反映出罗西的设计哲学：建筑应当是布莱希特戏剧的舞台布景，居住者可以在布景中自由地演绎自己的生活，"零度的"建筑形式可以充分激发居民的天然个性。加拉拉特西公寓是罗西关于建筑责任理论的综合阐释，同时，建筑师也在这项作品中重申了建筑形式中的各项最基本元素。

1969-1974

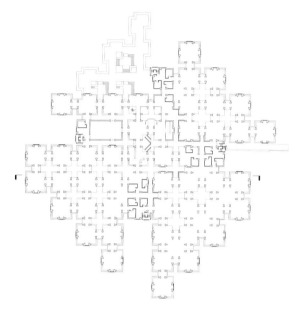

平面图

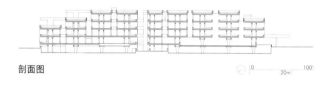

剖面图

CENTRAAL BEHEER BUILDING
HERMAN HERTZBERGER
Apeldoorn, The Netherlands

98

中央贝赫保险公司大楼
赫尔曼·赫茨贝格
荷兰，阿珀尔多伦
1968—1972 年

中央贝赫保险公司大楼开创了办公楼设计与工作场所文化的新轨迹，这栋建筑充分阐明了赫茨贝格的建筑理论，即每栋公共建筑都必须是一处"群落结构"（communal structure）——中央贝赫保险公司大楼实际上是一座小型的村庄，具备街道、广场与人群聚合的公共空间。这栋建筑共由 35 个 9 米 ×9 米的小型办公模块构成，单元之间虽然呈现聚合的态势却又彼此分离，保持着数米的距离。与此同时，模块化的结构允许这栋大楼根据功能需求灵活调整建筑布局与分区。中央贝赫保险公司大楼挑战了传统的空间等级制度，是"第十次小组"最具代表性的建筑作品之一（他们将每一个建筑都看作一片村庄）。在这栋建筑中，社交与互动行为是决定空间与功能配置的最终决定因素。

1968-1972

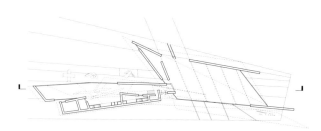

平面图

剖面图

北纬 47°36'01"

VITRA FIRE STATION
ZAHA HADID ARCHITECTS
Weil am Rhein, Germany

99

维特拉消防站
扎哈·哈迪德建筑师事务所

德国，莱茵河畔魏尔
1990—1993 年

  维特拉消防站是哈迪德首个建成的设计作品[1]，在这栋建筑中，哈迪德进行了一系列空间和形式实验，同时还将建筑与周围环境联系起来——至此，扎哈·哈迪德的绘画中体现出新构成主义语汇与动态和张力的相关理论终于实实在在地落在了地面上。狭长而低平的混凝土墙体构成了建筑的轮廓线，不断倾斜并折叠的构件之间表现出一种抽象的张力，造成了不稳定感；此外，不同的色彩与角度也造成了视觉上的误导效果，进一步增进了建筑的动态。这栋建筑还以纯粹而抽象的细节设计著称，各式各样的遮阳板和竖框遍布建筑内外。虽然维特拉消防站现在已被改造成一座展览馆，但它仍然是哈迪德职业生涯中最具争议性与可研究性的经典作品之一。

1990-1993

东经 7°36'52"

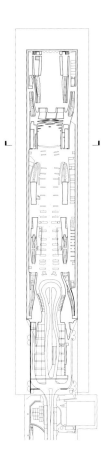

平面图

剖面图

北纬 35°27′04″

100

YOKOHAMA INTERNATIONAL PORT TERMINAL
FOREIGN OFFICE ARCHITECTS
Yokohama, Japan

横滨国际港客运中心
FOA 建筑师事务所

日本，横滨
1995—2002 年

横滨国际港客运中心最令人印象深刻的便是其顶部长达 430 米的流线型观景台：坡道、台阶、露台与种植了草坪的土丘布置得错落有致，别具一格的行走流线与单一流线方向的常规码头建筑形成了巨大反差，建筑师以多变的路径、各具特色的景观节点和不断变化的视野构成了其景观台的空间特征。在结构方面，客运中心主楼采用了一套以混凝土梁架为基础的混合结构系统，以两组平行的巨大支桩支起梁架，然后在梁架下方安装了金属折板。如此一来，不仅将结构的褶皱与缝隙变成了一种艺术化的折面形式，还调和统一了众多不同功能，如停车场、商店、餐馆、办公行政区和设备间的空间形式，连续统一的结构处理方式也大大增强了建筑抵抗地震横波的能力。横滨国际港客运中心的图纸看似简单，设计过程却极其复杂，是最早依赖全数字化计算与建模的建筑物之一，其结构和内部空间皆是一系列精心设计和测算的结果。

东经 139°38′50″

1995-2002

矩阵图

建筑师 / 建筑	萨伏伊别墅	朗香教堂	巴塞罗那德国馆	蓬皮杜艺术中心	约翰逊制蜡公司总部	范斯沃斯住宅	萨尔克生物研究所	玻璃之家	拉图雷特修道院	环球航空公司候机楼
阿部仁史	●	●	●	●	●	●	●	●		
斯坦·艾伦	●	●	●		●	●	●			
埃尔南·迪亚斯·阿隆索		●	●			●				
安藤忠雄	●	●	●		●	●	●	●	●	
坂茂	●	●	●	●	●	●	●	●		
马龙·布莱克韦尔	●	●	●	●	●	●	●	●		
亨利·N·科内	●	●	●	●	●	●	●			
普雷斯顿·斯科特·科恩		●	●		●	●	●			
彼得·库克爵士	●	●	●		●		●			
奥迪勒·德克		●	●		●		●			
尼尔·德纳里			●		●	●	●			
伊丽莎白·迪勒+里卡多·斯科菲迪奥+查尔斯·伦弗罗	●	●	●			●				
文卡·度别丹		●	●		●	●	●			
彼得·埃森曼	●	●	●		●	●	●			
珍妮·甘/Gang工作室	●	●	●	●	●	●	●			
扎哈·哈迪德女士	●	●	●		●	●	●			
克雷格·霍杰茨		●	●		●	●	●			
斯蒂文·霍尔	●	●	●		●	●	●			
弗朗辛·乌邦	●	●	●		●	●	●			
伊东丰雄		●	●			●				
金钟成		●	●		●	●	●			
莱昂·克里尔	●	●	●		●	●	●			
隈研吾	●	●	●	●	●	●	●	●		
拉斯·勒普		●	●			●				
丹尼尔·里伯斯金/里伯斯金工作室	●	●	●				●			
格雷格·林恩	●					●				
维尼·马斯+雅各布·范里斯+娜塔莉·德弗里斯	●	●	●	●	●	●	●			
槙文彦	●	●	●	●	●	●	●			
米夏埃尔·马尔灿	●	●	●		●					
汤姆·梅恩/墨菲西斯事务所	●	●	●	●	●	●	●			
理查德·迈耶	●	●	●		●	●	●			
拉斐尔·莫内欧	●	●	●			●				
埃里克·欧文·莫斯	●	●	●			●				
恩里克·诺尔滕/TEN事务所	●	●	●	●	●	●	●	●		
若泽·乌贝里		●					●			
西萨·佩里	●	●	●		●	●	●			
多米尼克·佩罗	●	●	●			●				
卡梅·皮诺斯		●	●			●				
詹姆斯·S·波尔舍克	●	●	●	●	●	●	●			
莫妮卡·庞塞·德莱昂		●	●			●				
安托万·普雷多克	●	●	●		●	●	●			
沃尔夫·D·普瑞/蓝天组	●	●	●			●				
杰西·赖泽+梅本奈奈子	●	●	●			●	●			
凯文·罗奇		●	●		●	●	●			
理查德·罗杰斯	●	●	●	●	●	●	●			
摩西·萨夫迪	●	●	●		●	●	●			
斯坦利·塞陶维兹	●	●	●		●	●	●			
麦克·斯科金	●	●	●		●	●	●			
妹岛和世+西泽立卫/SANNA事务所	●	●	●		●	●	●			
豪尔郝·西尔韦蒂		●	●			●	●			
安德烈娅·西米奇+瓦尔·K·沃克	●	●	●	●	●	●	●			
罗伯特·A·M·斯特恩	●	●	●		●	●	●			
安娜·托比通伊斯	●	●	●			●	●			
伯纳德·屈米	●	●	●			●				
本·范贝克尔+卡罗琳·博斯/UNS工作室	●	●	●	●	●	●	●			
罗伯特·文丘里+丹尼斯·斯科特·布朗	●	●	●			●	●			
马里昂·韦里+迈克尔·曼弗雷迪	●	●	●		●	●	●			
托德·威廉姆斯+比利·钱	●	●	●		●	●	●			

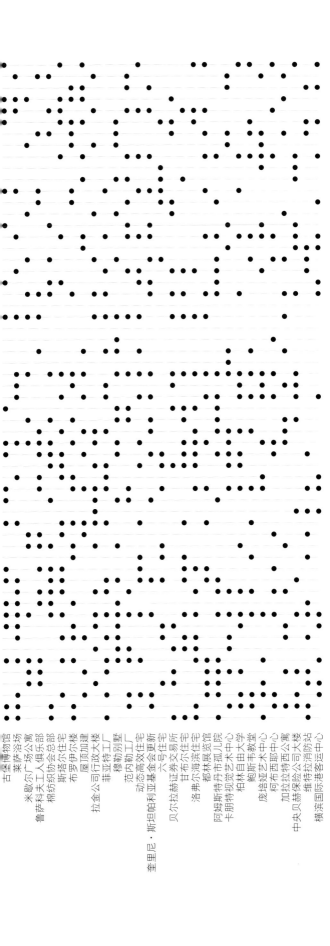

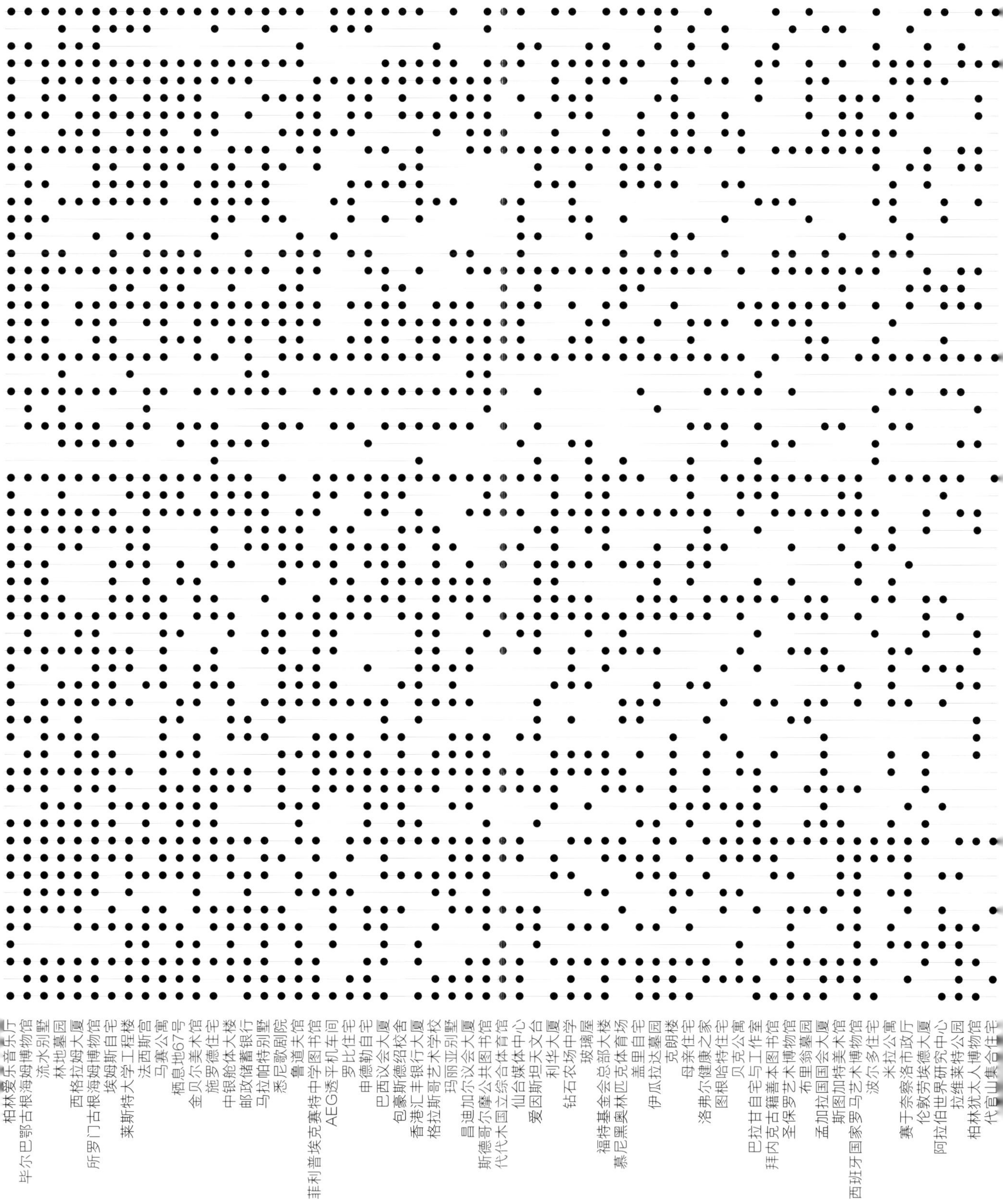

柏林爱乐音乐厅
毕尔巴鄂古根海姆博物馆
流水别墅
林地墓园
西格拉姆大厦
所罗门古根海姆博物馆
埃斯海姆博物馆
莱斯特大学工程楼
法西斯宫
马赛公寓
栖息地67号
施贝尔美德住宅
金贝尔美术馆
中银舱体大楼
邮政储蓄银行
马拉帕特别墅
悉尼歌剧院
鲁道夫美术馆
菲利普埃克赛特中学图书馆
AEG透平机车间
罗比住宅
巴塞勒会大厦
申西汇丰银行艺术学校舍
包豪斯绍根艺术学校
香港汇丰银行艺术学校
格拉斯哥艺术学校
玛丽亚别墅
昌迪加尔摩公共图书馆
斯德哥尔摩公共图书馆
代代木国立综合体育馆
仙台媒体中心
爱因斯坦天文台
钻石农场中学天台
利华大厦
玻璃屋
福特基金会总部大楼
慕尼黑奥林匹克体育场
盖蒂达墓园
伊瓜拉达墓园
克朗楼
母亲住宅
洛弗尔健康之家
巴拉甘自宅与工作室
拜内克古籍善本图书馆
圣保罗艺术博物馆
希里翁会大厦
孟加拉国家艺术博物馆
斯图加加德特美术馆
西班牙国家罗马艺术博物馆
波尔多住宅
米拉公寓
赛子奈察洛市政厅
伦敦劳埃德大厦
阿拉伯世界研究中心
拉维莱特公园
柏林犹太人博物馆
代官山集合住宅

建筑师的选择

阿部仁史
(Hitoshi Abe)

Thomas Edison Concrete House
AEG Turbine Factory
Frederick C. Robie House
Posen Tower
Postsparkasse
Schindler House
Schröderhuis
Lovell Beach House
Bauhaus Dessau
Karl-Marx-Hof
Lovell Health House
Stockholm Public Library
German Pavilion (Barcelona Pavilion)
Stockholm Exhibition of 1930
Villa Savoye
La Maison de Verre
Edgar J. Kaufmann House (Fallingwater)
Casa del Fascio
S.C. Johnson & Son Headquarters
Villa Mairea
The Woodland Cemetery
Casa Malaparte
Dymaxion House
Turin Exhibition Hall
The Eames House (Case Study House No. 8)
Kaufmann House
Farnsworth House
L'Unité d'Habitation
Lever House
Muuratsalo Experimental House
Iglesia de la Medalla Milagrosa
Chapelle Notre-Dame du Haut
Sydney Opera House
Crown Hall
Seagram Building
Philips Pavilion
National Congress, Brazil
Solomon R. Guggenheim Museum
Church of the Three Crosses
Stahl House (Case Study House No. 22)
Chemosphere
Couvent Sainte-Marie de La Tourette
TWA Flight Center
The Beinecke Rare Book & Manuscript Library
Nissay Theatre
Berliner Philharmonie
Leicester University Engineering Building
National Olympic Gymnasium, Tokyo
Kagawa Prefectural Government Hall
Rudolph Hall
Shrine of the Book
Centre Le Corbusier (Heidi Weber House)
Salk Institute
Sea Ranch Condominium Complex
Oita Prefectural Library
Kuwait Embassy, Tokyo
Sonsbeek Pavilion
Habitat '67
Ford Foundation Headquarters
Art Center College of Design, Pasadena
Hillside Terrace Complex I–VI
Nakagin Capsule Tower
Olympic Stadium, Munich
Kimbell Art Museum
Castelvecchio Museum
Free University of Berlin
Casa Bianchi
Museum of Modern Art, Gunma
Tanikawa House
Centraal Beheer Building
Bagsværd Church
Sumiyoshi Row House
World Trade Center Towers
Centre Pompidou
Brion Family Tomb
Gehry House
Teatro del Mondo
Kyungdong Presbyterian Church, Seoul
Tanimura Art Museum
Spiral, Tokyo
Steinhaus, Lake Ossiach
Lloyd's of London
Hongkong and Shanghai Bank Headquarters
Institut du Monde Arabe
Kate Mantilini, Beverly Hills
Tokyo Institute of Technology Centennial Hall
Rooftop Remodeling Falkestrasse
Igualada Cemetery
SYNTAX, Kyoto
Umeda Sky Building
UFA Cinema Centre
Maison à Bordeaux
Guggenheim Museum Bilbao
Jewish Museum Berlin
Expo 2000, Netherlands Pavilion
Sendai Mediatheque
Yokohama International Port Terminal
Diamond Ranch High School
Gifu Kitagata Apartment Building
Louis Vuitton Nagoya

斯坦·艾伦
(Stan Allen)

Ozenfant House
Villa Stein
Villa Savoye
Carpenter Center for the Visual Arts
Mill Owners' Association Building
Curutchet House
Couvent Sainte-Marie de La Tourette
Saint-Pierre, Firminy
Lange and Esters Houses
Tugendhat House
German Pavilion (Barcelona Pavilion)
Farnsworth House
New National Gallery
Norman Fisher House
First Unitarian Church of Rochester
National Assembly Building, Bangladesh
Franklin D. Roosevelt Four Freedoms Park
Studio Aalto
Säynätsalo Town Hall
Leicester University Engineering Building
Neue Staatsgalerie
Frederick C. Robie House
Larkin Company Administration Building
Solomon R. Guggenheim Museum
Stadtbad Mitte, Berlin
AEG Turbine Factory
Vanna Venturi House
Trubek and Wislocki Houses
National Gallery, Sainsbury Wing
Seattle Central Library
Plantahof Auditorium
DRLG Zentrale
Umlauftank
Barclay-Vesey Building
Commonwealth Building, Portland
Rusakov Workers' Club
Hunstanton School
The Economist Building, London
Casa Malaparte
Stockholm Public Library
The Woodland Cemetery
Woodland Chapel
Chapel of the Resurrection
Flower Kiosk, Malmö Cemetery
Hedmark Cathedral Museum
Nordic Pavillion, Venice
House I, Princeton
Berlin Memorial to the Murdered Jews of Europe
Benacerraf House Project
Town Hall, Logroño
Teatro del Mondo
Gallaratese II Apartments
Schröderhuis
Van Nelle Factory
PSFS Building
Arve Bridge
Gatti Wool Factory
La Maison de Verre
Robert Motherwell Quonset House
The Eames House (Case Study House No. 8)
Schindler House
Rockefeller Center
Municipal Bus Terminal, Salto
Bacardi Bottling Plant
Civil Government Building, Tarragona
Gimansio del Colegio Maravillas
Edificio de viviendas en la Barceloneta
Center for Hydrographic Studies, Madrid
Instituto de Nuestra Señora de la Victoria
Villa Bellotti, Milan
Casa al Parco
Casa del Fascio
Museum Insel Hombroich, Gallery Pavilions
Benedictine Monastery Chapel, Santiago
FAU, University of São Paulo
Museu de Arte de São Paulo
SESC Pompéia
Casa Butantã
Capela São Pedro Apóstolo
Museum of Sculpture, São Paulo
Igualada Cemetery
Torre Velasca
Padiglione d'Arte Contemporanea
Pirelli Tower
Gut Garkau Farm
Wall House II
Sky House, Tokyo
House with an Earthen Floor
White U
Tama Art University Library
21st Century Museum, Kanazawa
Piscinas de Marés
Iberê Camargo Foundation
Ricola Storage Building
House in Leymen
Laurenz-Stiftung Schaulager
Elbphilharmonie
Museo de Arte Contemporáneo de Castilla y León
Ron Davis House, Malibu
Frog Hollow

埃尔南·迪亚斯·阿隆索 (Hernan Diaz Alonso)

Igualada Cemetery
Barcelona Olympic Archery Range
Sports Center in Huesca
Scottish Parliament Building
Bank of London and South America
Villa La Roche
Curutchet House
Villa Savoye
Couvent Sainte-Marie de La Tourette
Saint-Pierre, Firminy
Carpenter Center for the Visual Arts
Chapelle Notre-Dame du Haut
Casa del Puente
Rooftop Remodeling Falkestrasse
Musée des Confluences
Louis Vuitton Foundation
Guggenheim Museum Bilbao
German Pavilion (Barcelona Pavilion)
Tugendhat House
New National Gallery
Farnsworth House
Glass House
Pantheon, Rome
Querini Stampalia Renovation
Quinta da Conceição, Swimming Pool
Maison à Bordeaux
The Woodland Cemetery
Mare de Déu de Montserrat de Montferri
Casa Iglesias, Barcelona
Church of Colònia Güell
Casa Batlló
Gimansio del Colegio Maravillas
Edificio Altamira
Église Sainte-Bernadette du Banlay
Einstein Tower
Jewelry Store Schullin
Zentralsparkasse Bank, Vienna
The Breuer Building (Whitney Museum)
Majolikahaus
Berlin Memorial to the Murdered Jews of Europe
San Carlo alle Quattro Fontane
Maison Coilliot
Mobius House
Wexner Center for the Visual Arts
Hôtel Tassel
Burgos Cathedral
Santa Maria del Mar, Barcelona
Vitra Fire Station
Ford Foundation Headquarters
The Beinecke Rare Book & Manuscript Library
Gardens of Versailles
Sir John Soane's Museum
Centre Pompidou
Lloyd's of London
Mosque–Cathedral of Córdoba
TWA Flight Center
S.C. Johnson & Son Headquarters
Palacio de los Deportes
Igreja da Pampulha
Walt Disney Concert Hall
Palais Garnier
MAXXI
Berliner Philharmonie
CCTV Headquarters
Rudolph Hall
Lippo Centre
RV (Room Vehicle) House Prototype
Water Pavilion H2O Expo
Vitra Design Museum
Beijing National Stadium
Sendai Mediatheque
41 Cooper Square
BMW Welt
Pterodactyl, Culver City
Villa dall'Ava
Emerson College Los Angeles
Church of Cristo Obrero
Villa Foscari
Stealth, Culver City
Torre Velasca
James R. Thompson Center
Vanna Venturi House
Van Nelle Factory
Karl-Marx-Hof
Glasgow School of Art
Post Office, Piazza Bologna
Teatro Regio
Solomon R. Guggenheim Museum
Robin Hood Gardens
History Faculty Building, Cambridge
Salk Institute
Elbphilharmonie
Mercedes-Benz Museum
Yokohama International Port Terminal
Parc de la Villette
Four Seasons Restaurant, Seagram Building
Seagram Building
Opéra Nouvel
Pérgolas en la Avenida de Icària
Storefront for Art and Architecture

安藤忠雄 (Tadao Ando)

Beurs van Berlage
Glasgow School of Art
Casa Milà
Michaelerplatz House (Looshaus)
Fagus Factory
Postsparkasse
Imperial Hotel, Tokyo
Stockholm City Hall
Église Notre-Dame du Raincy
Schröderhuis
Bauhaus Dessau
Sagrada Família
Stockholm Public Library
German Pavilion (Barcelona Pavilion)
Villa Savoye
La Maison de Verre
Edgar J. Kaufmann House (Fallingwater)
Casa del Fascio
Villa Mairea
The Woodland Cemetery
Casa Malaparte
Kaufmann House
Barragán House and Studio
Glass House
The Eames House (Case Study House No. 8)
Farnsworth House
Säynätsalo Town Hall
L'Unité d'Habitation
Lever House
Chapelle Notre-Dame du Haut
MIT Chapel
Sydney Opera House
Crown Hall
Otaniemi Chapel
Seagram Building
Louisiana Museum of Modern Art
National Congress, Brazil
Solomon R. Guggenheim Museum
Capilla de las Capuchinas
Stahl House (Case Study House No. 22)
Couvent Sainte-Marie de La Tourette
Halen Estate
TWA Flight Center
The Beinecke Rare Book & Manuscript Library
Berliner Philharmonie
Leicester University Engineering Building
National Olympic Gymnasium, Tokyo
Rudolph Hall
Vanna Venturi House
Salk Institute
Sea Ranch Condominium Complex
Piscinas de Marés
The Breuer Building (Whitney Museum)
Habitat '67
Smith House
Norman Fisher House
Hyatt Regency, San Francisco
Ford Foundation Headquarters
History Faculty Building, Cambridge
Cuadra San Cristóbal
Montreal Biosphère
Hillside Terrace Complex I–VI
Cubic Forest
Nakagin Capsule Tower
Olympic Stadium, Munich
Kimbell Art Museum
Gund Hall
Castelvecchio Museum
Douglas House
Marquette Plaza
Casa Bianchi
Museum of Modern Art, Gunma
Gallaratese II Apartments
Town Hall, Logroño
Marie Short House
La Fábrica, Sant Just Desvern
Sumiyoshi Row House
Sainsbury Center for Visual Arts
World Trade Center Towers
Centre Pompidou
Brion Family Tomb
Gehry House
Teatro del Mondo
Koshino House
Abteiberg Museum
Neue Staatsgalerie
Lloyd's of London
Hongkong and Shanghai Bank Headquarters
Church of the Light
The Menil Collection
Rooftop Remodeling Falkestrasse
Parc de la Villette
Goetz Collection
Maison à Bordeaux
Therme Vals
Kunsthaus Bregenz
Guggenheim Museum Bilbao
Diamond Ranch High School
Jewish Museum Berlin
Sendai Mediatheque

坂茂
(Shigeru Ban)

Larkin Company Administration Building
AEG Turbine Factory
Glasgow School of Art
Gamble House
Frederick C. Robie House
Casa Milà
Palais Stoclet
Postsparkasse
Schindler House
Ozenfant House
Stockholm City Hall
Van Nelle Factory
Schröderhuis
Lovell Beach House
Sagrada Família
Bauhaus Dessau
Villa Stein
Lovell Health House
Stockholm Public Library
German Pavilion (Barcelona Pavilion)
Tugendhat House
Villa Savoye
Paimio Sanatorium
La Maison de Verre
Edgar J. Kaufmann House (Fallingwater)
Casa del Fascio
Asilo Sant'Elia
S.C. Johnson & Son Headquarters
Villa Mairea
The Woodland Cemetery
Dymaxion House
Barragán House and Studio
Turin Exhibition Hall
Glass House
The Eames House (Case Study House No. 8)
Farnsworth House
Säynätsalo Town Hall
L'Unité d'Habitation
Lever House
Muuratsalo Experimental House
Gatti Wool Factory
Iglesia de la Medalla Milagrosa
Chapelle Notre-Dame du Haut
MIT Chapel
Villa Sarabhai
Crown Hall
Maisons Jaoul
Mill Owners' Association Building
Studio Aalto
Richards Medical Research Laboratories
Palazzetto dello Sport
Seagram Building
Philips Pavilion
National Congress, Brazil
Solomon R. Guggenheim Museum
Church of the Three Crosses
Los Manantiales
Capilla de las Capuchinas
Bacardi Bottling Plant
Stahl House (Case Study House No. 22)
Couvent Sainte-Marie de La Tourette
TWA Flight Center
Ena de Silva House
Querini Stampalia Renovation
Berliner Philharmonie
Leicester University Engineering Building
National Olympic Gymnasium, Tokyo
Otaniemi Technical University
Centre Le Corbusier (Heidi Weber House)
Salk Institute
The Assembly, Chandigarh
Piscinas de Marés
Edifício Copan
Habitat '67
Smith House
Norman Fisher House
Hyatt Regency, Atlanta
Ford Foundation Headquarters
New National Gallery
History Faculty Building, Cambridge
Kappe Residence
Cuadra San Cristóbal
Montreal Biosphère
Charles De Gaulle Airport, Terminal 1 & 2A–2F
Nakagin Capsule Tower
Olympic Stadium, Munich
Phillips Exeter Academy Library
Kimbell Art Museum
Castelvecchio Museum
Museum of Modern Art, Gunma
Sumiyoshi Row House
Centre Pompidou
Brion Family Tomb
Hedmark Cathedral Museum
Koshino House
National Assembly Building, Bangladesh
Lloyd's of London
Hongkong and Shanghai Bank Headquarters
Institut du Monde Arabe
Kunsthaus Bregenz

马龙·布莱克韦尔
(Marlon Blackwell)

Beurs van Berlage
Frederick C. Robie House
Fagus Factory
Merchants' National Bank, Grinnell
Schindler House
Sagrada Família
Lovell Beach House
Bauhaus Dessau
Stockholm Public Library
Villa Stein
German Pavilion (Barcelona Pavilion)
Villa Müller
Villa Savoye
Paimio Sanatorium
La Maison de Verre
Edgar J. Kaufmann House (Fallingwater)
Casa del Fascio
S.C. Johnson & Son Headquarters
Villa Mairea
Casa Malaparte
The Woodland Cemetery
Barragán House and Studio
The Eames House (Case Study House No. 8)
Glass House
Säynätsalo Town Hall
Farnsworth House
L'Unité d'Habitation
Lever House
Miller House, Columbus
MIT Chapel
The Assembly, Chandigarh
Chapelle Notre-Dame du Haut
Sydney Opera House
Crown Hall
Palazzetto dello Sport
Solomon R. Guggenheim Museum
National Library, Brasília
Haystack Mountain School of Crafts
Couvent Sainte-Marie de La Tourette
Capilla de las Capuchinas
Stahl House (Case Study House No. 22)
Prairie Chicken House
Esherick House
TWA Flight Center
Nordic Pavillion, Venice
Boa Nova Tea House
Berliner Philharmonie
Sheats Goldstein Residence
Rudolph Hall
Salk Institute
Sea Ranch Condominium Complex
The Breuer Building (Whitney Museum)
Habitat '67
Cuadra San Cristóbal
Kappe Residence
Flower Kiosk, Malmö Cemetery
Phillips Exeter Academy Library
Kimbell Art Museum
Castelvecchio Museum
San Cataldo Cemetery
Bagsværd Church
Centre Pompidou
Brion Family Tomb
Atheneum, New Harmony
Thorncrown Chapel
Magney House
Neue Staatsgalerie
Wexner Center for the Visual Arts
National Museum of Roman Art
Inn at Middleton Place
Ricola Storage Building
House VI
Stone House, Tavole
Saint Benedict Chapel, Sumvitg
Rooftop Remodeling Falkestrasse
Void Space/Hinged Space Housing
Igualada Cemetery
Goetz Collection
Kunsthal, Rotterdam
Signal Box Auf dem Wolf
Maison à Bordeaux
Vitra Fire Station
Yokohama International Port Terminal
Dominus Winery
Bruton Barr Library
Neurosciences Institute, La Jolla
Therme Vals
Burnette Residence
Kunsthaus Bregenz
Kew House, Victoria
Chapel of St. Ignatius
Palmer House, Tucson
Jewish Museum Berlin
Boyd Art Center, Riversdale
Diamond Ranch High School
Church of the Light
Keenan TowerHouse
Glass Chapel, Alabama
Gifu Kitagata Apartment Building
Sendai Mediatheque

亨利·N·科布
(Henry N. Cobb)

Flatiron Building
Beurs van Berlage
Darwin D. Martin House
Postsparkasse
Unity Temple
National Farmers Bank
Glasgow School of Art
Frederick C. Robie House
Casa Milà
New York Public Library
Palais Stoclet
Park Güell
Aircraft Hangars, Orly
The Cenotaph, Whitehall
Het Schip
Einstein Tower
Schindler House
L'Esprit Nouveau Pavilion
Schröderhuis
Bauhaus Dessau
Los Angeles Public Library
Villa Stein
Rusakov Workers' Club
Barclay-Vesey Building
Karl-Marx-Hof
Stockholm Public Library
German Pavilion (Barcelona Pavilion)
Villa Müller
Tugendhat House
Villa Savoye
PSFS Building
La Maison de Verre
Schwandbach Bridge
Rockefeller Center
Viipuri Municipal Library
Casa del Fascio
Göteborg Town Hall
S.C. Johnson & Son Headquarters
Edgar J. Kaufmann House (Fallingwater)
Villa Mairea
Chamberlain Cottage
Lloyd Lewis House
Turin Exhibition Hall
The Eames House (Case Study House No. 8)
Glass House
Casa Il Girasole
860–880 Lake Shore Drive
Säynätsalo Town Hall
Borsalino Housing
L'Unité d'Habitation
Coyoacán Market Hall
Chapelle Notre-Dame du Haut
National Pensions Building, Helsinki
Mill Owners' Association Building
Crown Hall
Richards Medical Research Laboratories
Edificio de viviendas en la Barceloneta
Seagram Building
Solomon R. Guggenheim Museum
McMath-Pierce Solar Telescope
The Assembly, Chandigarh
Carpenter Center for the Visual Arts
Berliner Philharmonie
Leicester University Engineering Building
Rudolph Hall
Vanna Venturi House
Salk Institute
The Breuer Building (Whitney Museum)
Fire Station No. 4, Columbus
Habitat '67
New National Gallery
Oakland Museum of California
Mummers Theater
House II (Vermont House)
Phillips Exeter Academy Library
Kimbell Art Museum
Douglas House
Gallaratese II Apartments
Whig Hall, Princeton University
Bagsværd Church
Centre Pompidou
Brion Family Tomb
Gehry House
Vietnam Veterans Memorial
Gordon Wu Hall
Neue Staatsgalerie
Loyola Law School Campus
Hongkong and Shanghai Bank Headquarters
SESC Pompéia
National Museum of Roman Art
Ricola Storage Building
Church of the Light
Kunsthal, Rotterdam
Vitra Fire Station
Neurosciences Institute, La Jolla
Chapel of St. Ignatius
Guggenheim Museum Bilbao
Murcia Town Hall
Diamond Ranch High School
Jewish Museum Berlin

普雷斯顿·斯科特·科恩
(Preston Scott Cohen)

Guaranty Building
National Farmers Bank
Union Bank, Columbus, Wisconsin
Peoples Savings Bank, Cedar Rapids
AEG Turbine Factory
Stockholm Public Library
Petersdorff Department Store
Barclay-Vesey Building
Flatiron Building
Downtown Athletic Club, New York
Glasgow School of Art
Larkin Company Administration Building
Frederick C. Robie House
Solomon R. Guggenheim Museum
S.C. Johnson & Son Headquarters
David and Gladys Wright House
Beth Sholom Congregation, Elkins Park
Rue Franklin Apartments
Villa Savoye
Villa Stein
Swiss Pavilion
Carpenter Center for the Visual Arts
The Assembly, Chandigarh
Couvent Sainte-Marie de La Tourette
L'Unité d'Habitation
Schröderhuis
Soviet Pavilion
Het Schip
Palais Stoclet
American Bar, Vienna
Villa Müller
Michaelerplatz House (Looshaus)
German Pavilion (Barcelona Pavilion)
Tugendhat House
Hubbe House
Seagram Building
Crown Hall
Farnsworth House
Inland Steel Building
Lever House
John Hancock Center
Casa Milà
Park Güell
Viipuri Municipal Library
Baker House
Aalto-Hochhaus, Bremen
Berliner Philharmonie
Palazzetto dello Sport
Saint Mary's Cathedral, San Francisco
San Juan de Ávila, Alcalá de Henares
The Breuer Building (Whitney Museum)
Centre Pompidou
The Pagoda, Madrid
Iglesia de la Medalla Milagrosa
Edificio de CEPAL
FAU, University of São Paulo
Nordic Pavillion, Venice
Sydney Opera House
SESC Pompéia
Richards Medical Research Laboratories
Kimbell Art Museum
Yale Center for British Art
Phillips Exeter Academy Library
National Assembly Building, Bangladesh
Rudolph Hall
Temple Street Parking Garage, New Haven
Boston City Hall
National Olympic Gymnasium, Tokyo
Marina City, Chicago
Casa del Fascio
Casa Il Girasole
CBS Building, New York
John Hancock Tower
Ford Foundation Headquarters
Hyatt Regency, Atlanta
Torre Velasca
Leicester University Engineering Building
Neue Staatsgalerie
AT&T Building, New York
Pennzoil Place
Vanna Venturi House
Brant House, Greenwich
Allen Memorial Art Museum, Oberlin
San Cataldo Cemetery
Wexner Center for the Visual Arts
Maison à Bordeaux
Zentrum für Kunst und Medientechnologie
Carlos Ramos Pavilion
Galician Center of Contemporary Art
Gehry House
Binoculars Building
Vitra Design Museum
Neuer Zollhof
National Museum of Roman Art
The Menil Collection
Signal Box Auf dem Wolf
Koechlin House
The Woodland Cemetery
House on a Curved Road
Bank of Georgia headquarters, Tbilisi

彼得・库克爵士
(Sir Peter Cook)

Melnikov House
Boots Pharmaceutical Factory
Samfunnshuset, Oslo
Bavinger House
Church of St. Michael, Frankfurt
Edificio Copan
Church of St. Peter, Klippan
Umlauftank
Tower of Winds, Yokohama
Spiral Apartment House, Ramat Gan
Shonandai Cultural Center
UFA Cinema Centre
Postsparkasse
Schindler House
Karl-Marx-Hof
Lovell Health House
Stockholm Public Library
Villa Moller
German Pavilion (Barcelona Pavilion)
Villa Savoye
Van Nelle Factory
La Maison de Verre
Villa Girasole
Edgar J. Kaufmann House (Fallingwater)
Casa del Fascio
The Woodland Cemetery
Casa Malaparte
Dymaxion House
Turin Exhibition Hall
Glass House
The Eames House (Case Study House No. 8)
Baker House
Muuratsalo Experimental House
Farnsworth House
L'Unité d'Habitation
Lever House
Crown Hall
Maisons Jaoul
Chapelle Notre-Dame du Haut
Seagram Building
National Congress, Brazil
Philips Pavilion
Solomon R. Guggenheim Museum
Cité d'Habitacion of Carrières Centrales
Stahl House (Case Study House No. 22)
Couvent Sainte-Marie de La Tourette
Municipal Orphanage, Amsterdam
Halen Estate
TWA Flight Center
Querini Stampalia Renovation
Berliner Philharmonie
Leicester University Engineering Building
Vanna Venturi House
National Olympic Gymnasium, Tokyo
Rudolph Hall
Salk Institute
The Assembly, Chandigarh
Centre Le Corbusier (Heidi Weber House)
Kuwait Embassy, Tokyo
Bank of London and South America
Habitat '67
Smith House
Museu de Arte de São Paulo
Ford Foundation Headquarters
Benacerraf House Project
Flower Kiosk, Malmö Cemetery
The Beinecke Rare Book & Manuscript Library
House III (Miller House)
Nakagin Capsule Tower
Olympic Stadium, Munich
Worker's Housing Estate, Hoek van Holland
Phillips Exeter Academy Library
Kimbell Art Museum
Free University of Berlin
Byker Wall
Centraal Beheer Building
Bagsværd Church
Brion Family Tomb
Centre Pompidou
SESC Pompéia
Hedmark Cathedral Museum
Gehry House
Koshino House
National Assembly Building, Bangladesh
Neue Staatsgalerie
Lloyd's of London
Hongkong and Shanghai Bank Headquarters
Rooftop Remodeling Falkestrasse
National Museum of Roman Art
Igualada Cemetery
Parc de la Villette
Maison à Bordeaux
Samitaur Tower
Guggenheim Museum Bilbao
Jewish Museum Berlin
Bordeaux Law Courts
Diamond Ranch High School
Yokohama International Port Terminal
Sendai Mediatheque
Einstein Tower

奥迪勒・德克
(Odile Decq)

Charles De Gaulle Airport, Terminal 1 & 2A–2F
Centre Pompidou
La Maison de Verre
Palazzetto dello Sport
German Pavilion (Barcelona Pavilion)
Casa del Fascio
French Communist Party Headquarters, Paris
Jewish Museum Berlin
UFA Cinema Centre
Berliner Philharmonie
Casa Malaparte
Villa Savoye
Maison Bernard
Maison Drusch
Zevaco Dome
Schröderhuis
Lloyd's of London
Robin Hood Gardens
Cave of Museum of Graffiti
Hundertwasserhaus
Nemausus
Narkomfin Building
Heinz-Galinski School
SAS Royal Hotel, Copenhagen
Torre Velasca
Berkeley City Club
Quartier de la Maladrerie, Aubervilliers
SESC Pompéia
London Stansted Airport
Habitat '67
Shrine of the Book
Bergiselschanze
Sydney Opera House
Eglise Sainte-Bernadette du Banlay
Climat de France, Algiers
University of Constantine
Hongkong and Shanghai Bank Headquarters
Banque Populaire de l'Ouest, Rennes
Nakagin Capsule Tower
Yamanashi Museum of Fruit
Olympic Stadium, Munich
American Bar, Vienna
Sheats Goldstein Residence
Atomium
Jewelry Store Schullin
Montreal Biosphère
Phillips Exeter Academy Library
Igualada Cemetery
Crematorium Baumschulenweg
Lord's Media Centre
Cathedral of Brasilia
Guggenheim Museum Bilbao
Solomon R. Guggenheim Museum
Parc de la Villette
New National Gallery
Pavilion for World Design Exposition, Nagoya
Bauhaus Dessau
UNESCO Headquarters, Paris
Pilgrimage Church, Neviges
The Eames House (Case Study House No. 8)
Einstein Tower
Maison Jean Prouvé, Nancy
Babylon Apartments, Miami
Marina City, Chicago
Balfron Tower
B 018
University of Urbino
Glass House
Bavinger House
Centraal Beheer Building
Maison d'Iran
Maison Bloc
L'Unité d'Habitation
Philips Pavilion
Alexandra Road Housing
Walden 7
Croulebarbe Tower
Amiraux Swimming Pool
Säynätsalo Town Hall
Shukhov Tower
Cardiff Bay Visitor Center
Signal Box Auf dem Wolf
Library Delft University of Technology
Bank of London and South America
Zolani Multi Purpose Centre
New Gourna Village Mosque
Ville évolutive, La Douvaine
Sawchu
Plastic House, Paris
Port la Galère
Gehry House
Montreal Biosphère
Maison Gonflable
House VI
E-1027
Maison à Bordeaux
Snowdon Aviary
Gratte-ciel, Villeurbanne
Melnikov House
Kunsthaus Graz

尼尔·德纳里 (Neil Denari)	伊丽莎白·迪勒 + 里卡多·斯科菲迪奥 + 查尔斯·伦弗罗 (Elizabeth Diller + Ricardo Scofidio + Charles Renfro)
Postsparkasse	Rue Franklin Apartments
Gamble House	AEG Turbine Factory
Casa Milà	Glasgow School of Art
Palais Stoclet	Casa Milà
Fagus Factory	Frederick C. Robie House
Frederick C. Robie House	Detroit Arsenal
Park Güell	Posen Tower
Het Schip	Postsparkasse
German Pavilion (Barcelona Pavilion)	Fiat Works
Schindler House	Shukhov Tower
Goetheanum	Schröderhuis
Einstein Tower	Villa Moller
Bauhaus Dessau	German Pavilion (Barcelona Pavilion)
Schröderhuis	Rusakov Workers' Club
Van Nelle Factory	Downtown Athletic Club, New York
Villa Cook	Salginatobel Bridge
La Maison de Verre	Villa Savoye
Tugendhat House	La Maison de Verre
Villa Müller	Haus Schminke
Villa Savoye	London Zoo Penguin Pool
Edgar J. Kaufmann House (Fallingwater)	Villa Girasole
Casa del Fascio	Casa del Fascio
Casa Malaparte	Merzbau
The Eames House (Case Study House No. 8)	S.C. Johnson & Son Headquarters
Baker House	Villa Mairea
Ariston Hotel, Mar del Plata	The Woodland Cemetery
Lever House	Casa Malaparte
Solomon R. Guggenheim Museum	Dymaxion House
Chapelle Notre-Dame du Haut	Barragán House and Studio
Farnsworth House	8x8 Demountable House
Ginásio do Clube Atlético Paulistano	Turin Exhibition Hall
Hunstanton School	The Eames House (Case Study House No. 8)
Crown Hall	Bavinger House
Richards Medical Research Laboratories	Farnsworth House
L'Unité d'Habitation	L'Unité d'Habitation
Studio Aalto	Das Canoas House
Sky House, Tokyo	Miller House, Columbus
National Congress, Brazil	Chapelle Notre-Dame du Haut
Olivetti Factory, Merlo	House of the Future, 1956
US Air Force Academy Cadet Chapel	Los Manantiales
Leicester University Engineering Building	Seagram Building
Salk Institute	Philips Pavilion
TWA Flight Center	Museum of Modern Art, Rio de Janeiro
Pepsi-Cola Building	Solomon R. Guggenheim Museum
Bank of London and South America	Church of the Three Crosses
Torre Blancas	Palácio do Planalto
Snowdon Aviary	Church of Cristo Obrero
Gwathmey Residence	Couvent Sainte-Marie de La Tourette
Edificio Copan	US Air Force Academy Cadet Chapel
Lieb House	TWA Flight Center
Fire Station No. 4, Columbus	Sheats Goldstein Residence
Kappe Residence	Gateway Arch
Smith House	Querini Stampalia Renovation
Balfron Tower	The Beinecke Rare Book & Manuscript Library
New National Gallery	Berliner Philharmonie
Rogers House	Leicester University Engineering Building
Wyss Garden Center	National Museum of Anthropology
Pilgrimage Church, Neviges	Chiesa dell'Autostrada del Sole
Olivetti Training Center	National Olympic Gymnasium, Tokyo
Nibankan	Shrine of the Book
The Pagoda, Madrid	The Pagoda, Madrid
Mummers Theater	Salk Institute
Zentralsparkasse Bank, Vienna	National Art Schools of Cuba
Yamanashi Press and Broadcasting Center	The Breuer Building (Whitney Museum)
Nakagin Capsule Tower	Église Sainte-Bernadette du Banlay
Olivetti Headquarters, Frankfurt	Piscinas de Marés
Centre Pompidou	Montessori School, Delft
White U	German Pavilion, Expo '67
Florey Building	Hyatt Regency, Atlanta
Kuwait Embassy, Tokyo	Villa Rosa, 1966–1970
House VI	Museu de Arte de São Paulo
Willis Faber and Dumas Headquarters	FAU, University of São Paulo
Kimbell Art Museum	Art Center College of Design, Pasadena
Olympic Stadium, Munich	Nakagin Capsule Tower
House on a Curved Road	Oase Nr. 7
Charles De Gaulle Airport, Terminal 1 & 2A–2F	Kimbell Art Museum
Bass Residence	Castelvecchio Museum
Gehry House	Free University of Berlin
Brant House, Greenwich	House VI
Centraal Beheer Building	Charles De Gaulle Airport, Terminal 1 & 2A–2F
Sainsbury Center for Visual Arts	Marina City, Chicago
Pennzoil Place	Bank of Georgia headquarters, Tbilisi
Rooftop Remodeling Falkestrasse	InterAction Centre, Kentish Town
El Helicoide	White U
Hongkong and Shanghai Bank Headquarters	World Trade Center Towers
Lloyd's of London	Centre Pompidou
Allied Bank Tower	Hedmark Cathedral Museum
Brühl Sports Center	Sitio Roberto Burle Marx
Venice III House	Hongkong and Shanghai Bank Headquarters
Vitra Design Museum	The Menil Collection
The Menil Collection	Institut du Monde Arabe
Guggenheim Museum Bilbao	Kunsthal, Rotterdam
Suzuki House	Maison à Bordeaux
Barcelona Olympic Archery Range	Therme Vals
Villa VPRO	Schouwburgplein
Diamond Ranch High School	Guggenheim Museum Bilbao
Miyagi Stadium	Diamond Ranch High School
Goetz Collection	Jewish Museum Berlin
Signal Box Auf dem Wolf	Sendai Mediatheque
Igualada Cemetery	Expo 2000, Netherlands Pavilion

文卡・度别丹
(Winka Dubbeldam)

TWA Flight Center
Alhambra, Granada
Community Centre, Bak
German Pavilion (Barcelona Pavilion)
Bass Residence
Rudolph Residence, 23 Beekman Place
Berlin State Library
Berliner Philharmonie
Beurs van Berlage
Le Saint James, Bouliac
Palácio da Alvorada
Petah Tikva
Brion Family Tomb
CaixaForum Madrid
California Academy of Sciences
Museo Canova
Capilla de las Capuchinas
Casa da Música
Casa Malaparte
Centre Pompidou
Chapelle Notre-Dame du Haut
Chiesa dell'Autostrada del Sole
Chrysler Building
Church of the Light
Glass Chapel, Alabama
Couvent Sainte-Marie de La Tourette
41 Cooper Square
Cuadra San Cristóbal
Das Canoas House
Diamond Ranch High School
Dominus Winery
Netherlands Embassy, Berlin
De Admirant Entrance Building
Edgar J. Kaufmann House (Fallingwater)
Farnsworth House
Saint-Pierre, Firminy
Glass House
Casa de Vidro
497 Greenwich Building
Guggenheim Museum Bilbao
High Line 23
Holy Spirit Church, Paks
Hongkong and Shanghai Bank Headquarters
House VI
Chapel of St. Ignatius
Ingalls Rink
Jean-Marie Tjibaou Cultural Centre
S.C. Johnson & Son Headquarters
Institut du Monde Arabe
Kiefhoek Housing Development
Casa Milà
Lincoln Center
La Maison de Verre
Marina City, Chicago
Sendai Mediatheque
Melnikov House
Mercedes-Benz Museum
Musée des Confluences
Yale Center for British Art
Nakagin Capsule Tower
Netherlands Dance Theatre, Hague
New National Gallery
Nine Bridges Country Club
No Mass House
NORD/LB Hanover
Olympic Stadium, Munich
Municipal Orphanage
Parco Acqua Santa
Parco della Musica
Perot Museum of Nature and Science
Phaeno Science Center
Potala Palace Renovation
Prada Aoyama
Reichstag Dome
Schröderhuis
Frederick C. Robie House
Yokohama International Port Terminal
Rusakov Workers' Club
Sagrada Família
Salk Institute
Zonnenstraal
Seattle Central Library
History Faculty Building, Cambridge
Serpentine Gallery Pavilion, 2002
Sydney Opera House
Sonneveld House
Spanish Pavilion at Shanghai Expo 2010
Steinhaus, Lake Ossiach
Tate Modern
Igreja da Pampulha
Sheats Goldstein Residence
Therme Vals
National Congress, Brazil
Van Nelle Factory
Villa Savoye
Tugendhat House
Walt Disney Concert Hall
The Breuer Building (Whitney Museum)
Vitra Design Museum
520 West 28th Street

彼得・埃森曼
(Peter Eisenman)

Palais de la musique et des congrès
Maison Dom-ino
Maison Citrohan
Villa Savoye
Villa Stein
Swiss Pavilion
City of Refuge, Paris
Couvent Sainte-Marie de La Tourette
German Pavilion (Barcelona Pavilion)
Brick Country House
New National Gallery
Crown Hall
Farnsworth House
Villa Müller
Villa Moller
Michaelerplatz House (Looshaus)
Larkin Company Administration Building
Darwin D. Martin House
Solomon R. Guggenheim Museum
Palicka Villa
Basel Peterschule
Zonnenstraal
Werkbundsiedlung, Vienna
Berliner Philharmonie
Proposal for the Bank of Dresden
Casa del Fascio
Casa Giuliani Frigerio
Asilo Sant'Elia
Casa Cattaneo a Cernobbio
Fencing Academy, Rome
Casa Il Girasole
Corso Italia Complex, Milan
Casa Rustici
Istituto Marchiondi Spagliardi
AEG Turbine Factory
Goetheanum
Einstein Tower
Narkomfin Building
History Faculty Building, Cambridge
Leicester University Engineering Building
Florey Building
Neue Staatsgalerie
Braun AG Headquarters, Melsungen
Enschede Dormitory
San Cataldo Cemetery
Gallaratese II Apartments
The Economist Building, London
Golden Lane Housing
Robin Hood Gardens
Vanna Venturi House
Yale Mathematics Building Competition 1970
National Gallery, Sainsbury Wing
College Football Hall of Fame Proposal, VSBA
Smith House
Hanselmann House
Benacerraf House Project
Snyderman House
Whitney Museum Expansion Proposal, M. Graves
Fargo-Moorhead Cultural Center Bridge
Gwathmey Residence
Berlin Masque
Wall House II
House II (Vermont House)
Berlin Memorial to the Murdered Jews of Europe
Wexner Center for the Visual Arts
Design, Architecture, Art & Planning Building
Rudolph Hall
Sarasota High School
Glass House
Kline Biology Tower
Léon Krier (overall oeuvre)
Town Hall, Logroño
Bankinter Building
Bouça Saal Housing Complex
Jewelry Store Schullin
Museum of Modern Art, Gunma
Naoshima Contemporary Art Museum
Sendai Mediatheque
Austrian Cultural Forum
Boston City Hall
Miller House, Columbus
Knights of Columbus Building
Pacific Design Center
Très Grande Bibliothèque
Jussieu – Two Libraries
Parc de La Villette, OMA proposal
Eurodisney, Marne-la-Vallée, OMA proposal
Peak Leisure Club
Rosenthal Center for Contemporary Art
Herta and Paul Amir Building
6th Street House, Santa Monica
2-4-6-8 House
Chapel of St. Ignatius
Peter B. Lewis Building, CWRU
Guggenheim Museum Bilbao
Ron Davis House, Malibu
Schröderhuis
Beurs van Berlage
Säynätsalo Town Hall
Tallinn Art Museum Proposal

珍妮·甘 / Gang 工作室
(Jeanne Gang / Studio Gang)

SESC Pompéia
S.C. Johnson & Son Headquarters
Sydney Opera House
German Pavilion (Barcelona Pavilion)
Crown Hall
Sagrada Família
Chapelle Notre-Dame du Haut
Phillips Exeter Academy Library
Yale Center for British Art
The Beinecke Rare Book & Manuscript Library
Casa de Vidro
Villa Mairea
Tugendhat House
Casa del Fascio
University of Virginia
Viipuri Municipal Library
Church of Cristo Obrero
Centraal Beheer Building
Querini Stampalia Renovation
National Assembly Building, Bangladesh
Salk Institute
Edgar J. Kaufmann House (Fallingwater)
Centre Pompidou
La Maison de Verre
Louisiana Museum of Modern Art
Millard House
Berliner Philharmonie
Olympic Stadium, Munich
Habitat '67
E-1027
The Eames House (Case Study House No. 8)
Farnsworth House
New National Gallery
Frederick C. Robie House
Göteborg Town Hall
Park Güell
Santuario Dom Bosco
TWA Flight Center
Mill Owners' Association Building
Van Nelle Factory
The Assembly, Chandigarh
Casa Milà
Flatiron Building
Castelvecchio Museum
Montreal Biosphère
Auditorium Building, Chicago
Finlandia Hall
Kimbell Art Museum
Casa Butantã
Rudolph Hall
Villa Savoye
Museu de Arte de São Paulo
Säynätsalo Town Hall
Unity Temple
Schindler House
Indian Institute of Management, Ahmedabad
Edifício Jaraguá
Cathedral of Brasília
Itamaraty Palace
Postsparkasse
Ford Motor Company Glass Manufacturing Plant
Couvent Sainte-Marie de La Tourette
Carpenter Center for the Visual Arts
Gateway Arch
Bibliothèque nationale de France
Torres del Parque
Bulgwang-dong Catholic Church
Sir John Soane's Museum
Apolloscholen
National Art Schools of Cuba
Edificio Copan
FAU, University of São Paulo
Marquette Building
L'Unité d'Habitation
Sangath
The Woodland Cemetery
Carson, Pirie, Scott and Company Building
Studio Aalto
Palazzetto dello Sport
Monticello
Neutra Research House
Pantheon, Rome
Chand Baori
Notre Dame de Paris
Igualada Cemetery
Elbphilharmonie
MAXXI
Casa da Música
Beijing Airport
The Menil Collection
Universita Luigi Bocconi
Heydar Aliyev Center
UC Campus Recreation Center
Berlin Memorial to the Murdered Jews of Europe
Maison à Bordeaux
Ricola Storage Building
Louis Vuitton Foundation
Tama Art University Library
Sendai Mediatheque
Nexus World Housing

扎哈·哈迪德女士
(Dame Zaha Hadid)

American Bar, Vienna
Einstein Tower
Zuev Workers' Club
Svoboda Factory Club
Melnikov House
Nikolaev's House
London Zoo Penguin Pool
Highpoint Apartment Blocks
Finsbury Health Centre
BLPS Housing
Rockefeller Center
Gustavo Capanema Palace
Igreja da Pampulha
Boavista Bank Headquarters, Rio de Janeiro
Royal Festival Hall
Das Canoas House
Usk Street Estate
UNESCO Headquarters, Paris
Marina City, Chicago
Royal College of Physicians
Trellick Tower
Willis Faber and Dumas Headquarters
Royal National Theatre
Casa Gilardi
Sainsbury Center for Visual Arts
Qatar University
Vitra Campus
National Library of the Argentine Republic
Lord's Media Centre
Postsparkasse
Schindler House
Fiat Works
Schröderhuis
Cité Frugès Housing Complex
Karl-Marx-Hof
Lovell Health House
Stockholm Public Library
Villa Moller
German Pavilion (Barcelona Pavilion)
Villa Savoye
Van Nelle Factory
La Maison de Verre
Villa Girasole
Edgar J. Kaufmann House (Fallingwater)
Casa del Fascio
The Woodland Cemetery
Casa Malaparte
Dymaxion House
Turin Exhibition Hall
Glass House
The Eames House (Case Study House No. 8)
Baker House
Muuratsalo Experimental House
Farnsworth House
L'Unité d'Habitation
Lever House
Pedregulho Housing Complex
Crown Hall
Maisons Jaoul
Chapelle Notre-Dame du Haut
Seagram Building
National Congress, Brazil
Philips Pavilion
Solomon R. Guggenheim Museum
Cité d'Habitacion of Carrières Centrales
Stahl House (Case Study House No. 22)
Couvent Sainte-Marie de La Tourette
Municipal Orphanage
Halen Estate
TWA Flight Center
Querini Stampalia Renovation
Berliner Philharmonie
Leicester University Engineering Building
Vanna Venturi House
National Olympic Gymnasium, Tokyo
Peabody Terrace
Rudolph Hall
Salk Institute
The Assembly, Chandigarh
Centre Le Corbusier (Heidi Weber House)
Piscinas de Marés
Kuwait Embassy, Tokyo
Bank of London and South America
Habitat '67
Smith House
Museu de Arte de São Paulo
Ford Foundation Headquarters
Kowloon Walled City
Benacerraf House Project
Flower Kiosk, Malmö Cemetery
Petrobras Headquarters
The Beinecke Rare Book & Manuscript Library
House III (Miller House)
Nakagin Capsule Tower
Olympic Stadium, Munich
Phillips Exeter Academy Library
Kimbell Art Museum
Free University of Berlin
Byker Wall
Centraal Beheer Building

扎哈·哈迪德女士（续）

Whig Hall, Princeton University
Bagsværd Church
Brion Family Tomb
Centre Pompidou
SESC Pompéia
Hedmark Cathedral Museum
Gehry House
Atheneum, New Harmony
Koshino House
Vietnam Veterans Memorial
National Assembly Building, Bangladesh
Neue Staatsgalerie
Lloyd's of London
Hongkong and Shanghai Bank Headquarters
Rooftop Remodeling Falkestrasse
National Museum of Roman Art
Void Space/Hinged Space Housing
Igualada Cemetery
Parc de la Villette
Maison à Bordeaux
Samitaur Tower
Guggenheim Museum Bilbao
UFA Cinema Centre
Jewish Museum Berlin
Bordeaux Law Courts
Diamond Ranch High School
Yokohama International Port Terminal
Sendai Mediatheque

克雷格·霍杰茨
(Craig Hodgetts)

TWA Flight Center
General Motors Technical Center
Washington Dulles International Airport
MIT Chapell.
Rudolph Hall
Dipoli
Säynätsalo Town Hall
Riola Parish Church
Otaniemi Technical University
Maison du Brésil
Chapelle Notre-Dame du Haut
L'Unité d'Habitation
City of Refuge, Paris
Philips Pavilion
Couvent Sainte-Marie de La Tourette
Carpenter Center for the Visual Arts
History Faculty Building, Cambridge
Florey Building
Southgate Estate
Neue Staatsgalerie
Leicester University Engineering Building
Bank of London and South America
Berlin State Library
Berliner Philharmonie
Renault Distribution Centre
Centre Pompidou
Lloyd's of London
Richards Medical Research Laboratories
Centraal Beheer Building
Expo '70 Osaka, Festival Plaza
National Olympic Gymnasium, Tokyo
Habitat '67
2-4-6-8 House
UC Campus Recreation Center
41 Cooper Square
Santa Caterina Market
Scottish Parliament Building
Chrysler Building
PSFS Building
Rockefeller Center
United Nations Headquarters
Cincinnati Union Terminal Station
The Menil Collection
The Shard
Nakagin Capsule Tower
Maison Tropicale
Taliesin West
S.C. Johnson & Son Headquarters
Edgar J. Kaufmann House (Fallingwater)
Towell Library
Vitra Fire Station
Rooftop Remodeling Falkestrasse
Groninger Museum
The Eames House (Case Study House No. 8)
Pepsi-Cola Building
Inland Steel Building
Blur Building
Bibliotheca Alexandrina
Prada Aoyama
Dominus Winery
Signal Box Auf dem Wolf
Linked Hybrid
Chapel of St. Ignatius
Nelson-Atkins Museum of Art Addition
Horton Plaza
Canal City Hakata, Fukuoka
Olympic Stadium, Munich
German Pavilion, Expo '67
Seattle Central Library
House in Leymen
Guggenheim Museum Bilbao
Gehry House
Binoculars Building
Neutra Research House
Reiner-Burchill Residence
El Pueblo Ribera
High Line 23
Bauhaus Dessau
La Maison de Verre
Marie Short House
Einstein Tower
Futurama, 1939 New York World's Fair
National Congress, Brazil
Edificio Copan
Mercado de Colón
St. John's Abbey, Collegeville
IBM France
Palazzetto dello Sport
Orvieto Aircraft Hangars
Glass Pavilion
Sydney Opera House
Chiesa dell'Autostrada del Sole
Hayden Tract
Parc de la Villette
Le Fresnoy Art Center
Church on the Water
Yokohama International Port Terminal
Institut du Monde Arabe
Mimesis Museum
Gare do Oriente

斯蒂文·霍尔
(Steven Holl)

Glasgow School of Art
Larkin Company Administration Building
Unity Temple
Frederick C. Robie House
Steiner House
Postsparkasse
Michaelerplatz House (Looshaus)
Merchants' National Bank, Grinnell
Imperial Hotel, Tokyo
Einstein Tower
Schindler House
Ozenfant House
Villa La Roche
Schröderhuis
Tristan Tzara House
Cranbrook Academy of Art
Bauhaus Dessau
Villa Moller
Melnikov House
German Pavilion (Barcelona Pavilion)
Villa Müller
Villa Savoye
La Maison de Verre
Rockefeller Center
Edgar J. Kaufmann House (Fallingwater)
Casa del Fascio
S.C. Johnson & Son Headquarters
Casa Malaparte
Villa Mairea
The Eames House (Case Study House No. 8)
Farnsworth House
L'Unité d'Habitation
Baker House
Barragán House and Studio
Chapelle du Rosaire de Vence
Säynätsalo Town Hall
Crown Hall
Mill Owners' Association Building
The Assembly, Chandigarh
Chapelle Notre-Dame du Haut
House of Culture
St. Mark's Church in Björkhagen
Municipal Orphanage
National Congress, Brazil
Berliner Philharmonie
Torres de Satélite
Church of the Three Crosses
Couvent Sainte-Marie de La Tourette
Richards Medical Research Laboratories
Sydney Opera House
Seagram Building
Philips Pavilion
Rudolph Hall
Museu de Arte de São Paulo
Salk Institute
Solomon R. Guggenheim Museum
Leicester University Engineering Building
TWA Flight Center
National Assembly Building, Bangladesh
Indian Institute of Management, Ahmedabad
Querini Stampalia Renovation
The Beinecke Rare Book & Manuscript Library
Ford Foundation Headquarters
National Olympic Gymnasium, Tokyo
History Faculty Building, Cambridge
Phillips Exeter Academy Library
Church of St. Peter, Klippan
Centre Le Corbusier (Heidi Weber House)
Habitat '67
Hanselmann House
Smith House
Gallaratese II Apartments
Olympic Stadium, Munich
Bagsværd Church
Brion Family Tomb
Kimbell Art Museum
Nakagin Capsule Tower
Museum of Modern Art, Gunma
Centre Pompidou
House VI
San Cataldo Cemetery
Beires House
Atheneum, New Harmony
Uehara House
Neue Staatsgalerie
Institut du Monde Arabe
The Menil Collection
Parc de la Villette
Tokyo Institute of Technology Centennial Hall
Faculty of Architecture of the University of Porto
Jewish Museum Berlin
Church of the Light
Void Space/Hinged Space Housing
Norwegian Glacier Museum
Vitra Fire Station
Maison à Bordeaux
UFA Cinema Centre
Guggenheim Museum Bilbao
Diamond Ranch High School
Sendai Mediatheque

弗朗辛·乌邦
(Francine Houben)

Villa Savoye
L'Unité d'Habitation
Chapelle Notre-Dame du Haut
The Assembly, Chandigarh
Couvent Sainte-Marie de La Tourette
Edgar J. Kaufmann House (Fallingwater)
S.C. Johnson & Son Headquarters
Taliesin West
Jacobs First House
Solomon R. Guggenheim Museum
Villa Mairea
Villa Moller
Das Canoas House
National Congress, Brazil
Ibirapuera Park
Casa de Vidro
SESC Pompéia
Museu de Arte de São Paulo
Sydney Opera House
Saint Catherine's College, Oxford
Old Faithful Inn
Farnsworth House
German Pavilion (Barcelona Pavilion)
Seagram Building
Tugendhat House
The Eames House (Case Study House No. 8)
Bailey House (Case Study House No. 21)
Salk Institute
Phillips Exeter Academy Library
Kimbell Art Museum
Rudolph Hall
Dymaxion House
National Olympic Gymnasium, Tokyo
Nakagin Capsule Tower
Igualada Cemetery
Sagrada Família
Casa Milà
Bauhaus Dessau
TWA Flight Center
Gateway Arch
UNESCO Headquarters, Paris
Schindler House
Lovell Health House
Lovell Beach House
Gamble House
Chemosphere
Kaufmann House
Van Nelle Factory
Vanna Venturi House
Centre Pompidou
Ford Foundation Headquarters
National Gallery of Art, East Building
Haus Schminke
Berliner Philharmonie
Hyatt Regency, Atlanta
The Beinecke Rare Book & Manuscript Library
La Maison de Verre
Zonnenstraal
Beurs van Berlage
Het Schip
Postsparkasse
Barragán House and Studio
Glasgow School of Art
Stockholm Public Library
Hilversum Town Hall
Casa del Fascio
Vitra Fire Station
Kunsthal, Rotterdam
Library Delft University of Technology
Piscinas de Marés
Therme Vals
Chrysler Building
Empire State Building
World Trade Center Towers
Melnikov House
Jean Prouvé Housing Elements & Façades
Brion Family Tomb
Municipal Museum, The Hague
Guggenheim Museum Bilbao
John Hancock Center
Casa Malaparte
Glass House
Silver Hut
Sendai Mediatheque
Hongkong and Shanghai Bank Headquarters
Tate Modern
Royal Festival Hall
Kanchanjunga Apartments
Hufeisensiedlung
Papaverhof
Highpoint Apartment Blocks
Huis Van der Leeuw
Schröderhuis
Fiat Works
Café De Unie
Auditorium of the Technical University in Delft
Kasbah Housing, Hengelo
Municipal Orphanage, Amsterdam
Mobius House
Villa VPRO

伊东丰雄
(Toyo Ito)

Postsparkasse
Villa Savoye
The Eames House (Case Study House No. 8)
L'Unité d'Habitation
Chapelle Notre-Dame du Haut
Sydney Opera House
Villa Sarabhai
National Congress, Brazil
Church of the Three Crosses
Couvent Sainte-Marie de La Tourette
Municipal Orphanage
TWA Flight Center
Berliner Philharmonie
National Olympic Gymnasium, Tokyo
Vanna Venturi House
Salk Institute
The Assembly, Chandigarh
Glasgow School of Art
Frederick C. Robie House
Schindler House
Stockholm Public Library
Villa Moller
German Pavilion (Barcelona Pavilion)
Edgar J. Kaufmann House (Fallingwater)
S.C. Johnson & Son Headquarters
Villa Mairea
Dymaxion House
Glass House
Farnsworth House
Mill Owners' Association Building
Solomon R. Guggenheim Museum
Stahl House (Case Study House No. 22)
Toulouse-le-Mirail Housing
Halen Estate
Querini Stampalia Renovation
John Deere World Headquarters
Otaniemi Technical University
Rudolph Hall
Sea Ranch Condominium Complex
Habitat '67
Montreal Biosphère
Nakagin Capsule Tower
Olympic Stadium, Munich
Kimbell Art Museum
Castelvecchio Museum
Museum of Modern Art, Gunma
World Trade Center Towers
Centre Pompidou
Brion Family Tomb
Gehry House
Teatro del Mondo
Sumiyoshi Row House
National Assembly Building, Bangladesh
Hongkong and Shanghai Bank Headquarters
Institut du Monde Arabe
Void Space/Hinged Space Housing
Maison à Bordeaux
Guggenheim Museum Bilbao
Diamond Ranch High School
Sendai Mediatheque
Yokohama International Port Terminal

金正先
(Jong Soung Kimm)

Beurs van Berlage
Postsparkasse
Glasgow School of Art
Rue Franklin Apartments
Gamble House
AEG Turbine Factory
Michaelerplatz House (Looshaus)
Carson, Pirie, Scott and Company Building
Larkin Company Administration Building
Frederick C. Robie House
Edgar J. Kaufmann House (Fallingwater)
S.C. Johnson & Son Headquarters
Unity Temple
Solomon R. Guggenheim Museum
German Pavilion (Barcelona Pavilion)
Tugendhat House
860—880 Lake Shore Drive
Farnsworth House
Crown Hall
Seagram Building
New National Gallery
Villa Stein
Villa Savoye
L'Unité d'Habitation
Chapelle Notre-Dame du Haut
Couvent Sainte-Marie de La Tourette
The Assembly, Chandigarh
Mill Owners' Association Building
Einstein Tower
Bauhaus Dessau
La Maison de Verre
Sagrada Família
Park Güell
Casa Milà
Casa del Fascio
Schröderhuis
Stockholm Public Library
The Woodland Cemetery
St. Mark's Church in Björkhagen
Fiat Works
Schindler House
Lovell Health House
Rusakov Workers' Club
Säynätsalo Town Hall
Baker House
The Eames House (Case Study House No. 8)
Richards Medical Research Laboratories
Salk Institute
Kimbell Art Museum
Phillips Exeter Academy Library
Yale Center for British Art
National Assembly Building, Bangladesh
Lever House
The Beinecke Rare Book & Manuscript Library
Glass House
Rudolph Hall
Sydney Opera House
Berliner Philharmonie
National Congress, Brazil
Leicester University Engineering Building
TWA Flight Center
John Deere World Headquarters
Ford Foundation Headquarters
Vanna Venturi House
Querini Stampalia Renovation
Brion Family Tomb
Palazzo del Lavoro
Turin Exhibition Hall
Free University of Berlin
National Olympic Gymnasium, Tokyo
The Breuer Building (Whitney Museum)
Hillside Terrace Complex I–VI
Centraal Beheer Building
San Cataldo Cemetery
Gallaratese II Apartments
Municipal Orphanage
National Gallery of Art, East Building
Pyramide du Louvre
Bank of China Tower, Hong Kong
Richard J. Daley Center
Sears Tower
Smith House
Getty Center
Olympic Stadium, Munich
Centre Pompidou
The Menil Collection
Gehry House
Guggenheim Museum Bilbao
Hongkong and Shanghai Bank Headquarters
Lloyd's of London
Parc de la Villette
National Museum of Roman Art
Institut du Monde Arabe
Jewish Museum Berlin
Yokohama International Port Terminal
Igualada Cemetery
Maison à Bordeaux
Naoshima Contemporary Art Museum
Sendai Mediatheque
Diamond Ranch High School

莱昂·克里尔
（Léon Krier）

Kirche am Steinhof
Postsparkasse
Pan American Union Building
The House of the Temple
Sanatorium Purkersdorf
Palais Stoclet
Panama–California Exposition
Colonial Williamsburg Reconstructions
Hydro-Electric Plant, Riva del Garda
Vittoriale degli italiani
Hydroelectric Plant, Pienza
Prague Castle Interventions
Lincoln Memorial
Zeppelinfeld
Villa Savoye
Midland Bank, Manchester
Rue Mallet-Stevens
Siedlung Margarethenhöhe
Palazzo della Civiltà Italiana
Palazzo dei Congressi
Villa Bianca, Seveso
Firenze Santa Maria Novella
Kallithea Springs
Arco della Vittoria
Garbatella Quarter
French Embassy, Belgrade
The Woodland Cemetery
Lister County Courthouse, Sölvesborg
Copenhagen Police Headquarters
Faaborg Museum
Hilversum Town Hall
Villa Becker, Munich
Königsplatz, Munich
Texas Centennial Exposition
Chapelle Notre-Dame du Haut
The Assembly, Chandigarh
Plaza de España, Seville
Ibero-American Exposition of 1929
Petit Palais
Grand Palais
Pont Alexandre III
Aula Palatina
Villa Kerylos
Ise Grand Shrine
Davenport and Pierson Colleges, Yale
Hoover Dam
La Maison de Verre
Zonnestraal
Leicester University Engineering Building
Pasadena City Hall
Charles De Gaulle Airport, Terminal 1 & 2A–2F
Baikonur Cosmodrome Launch Site 250
Miami Biltmore Hotel
Grand Central Terminal
Pennsylvania Station
Union Station, Washington, D.C.
Adolphe Bridge
Eccles Building
Villa Wagner II
Casa Malaparte
Reconstruction of Old Warsaw
Frauenkirche Dresden Reconstruction
German Pavilion (Barcelona Pavilion)
Windsor, Vero Beach
Seaside Chapel, Ruskin Square
Flak Towers, Vienna
Golden Gate Bridge
Festspielhaus Hellerau
S.C. Johnson & Son Headquarters
Celebration Masterplan
Il villaggio Cesare Battisti
All-Union Agricultural Exhibition
New Gourna Village Mosque
Island Mosque, Jeddah
Via Roma, Torino
1939 Reconstruction of Devastated Regions, Spain
Kongresshalle
Hoover Building
Detroit Arsenal
De La Warr Pavilion
Halen Estate
Barragán House and Studio
TWA Flight Center
Palm Beach Town Hall Square
San Francisco Maritime Museum
Richmond Riverside Development
Grundtvig Memorial Church
Vilanova House, Key Biscayne
Stockholm Public Library
New Reich Chancellery
Plečnik Žale Cemetery
Ljubljana Central Market
Triple Bridge
Søndermark Chapel and Crematorium
Øregård Gymnasium
Atlantic Wall Bunkers & Towers
McMillan Plan
Stoaken Garden City, Berlin
Foro Italico
Port Grimaud

隈研吾
（Kengo Kuma）

Métropolitain Entrance
Beurs van Berlage
Rue Franklin Apartments
Larkin Company Administration Building
AEG Turbine Factory
Glasgow School of Art
Gamble House
Unity Temple
Frederick C. Robie House
Casa Milà
Palais Stoclet
Posen Tower
Fagus Factory
Postsparkasse
Park Güell
Het Schip
Einstein Tower
Schindler House
Fiat Works
Monument to the March Dead
Stockholm City Hall
Van Nelle Factory
Schröderhuis
Café De Unie
Lovell Beach House
Schocken Department Store, Stuttgart
Sagrada Família
Bauhaus Dessau
Barclay-Vesey Building
Karl-Marx-Hof
Lovell Health House
Stockholm Public Library
Villa Moller
E-1027
Goetheanum
German Pavilion (Barcelona Pavilion)
Rusakov Workers' Club
Narkomfin Building
Stockholm Exhibition of 1930
Villa Müller
Tugendhat House
Villa Savoye
Hilversum Town Hall
Paimio Sanatorium
La Maison de Verre
Haus Schminke
Villa Girasole
Edgar J. Kaufmann House (Fallingwater)
Casa del Fascio
Michaelerplatz House (Looshaus)
Asilo Sant'Elia
S.C. Johnson & Son Headquarters
Villa Mairea
Arve Bridge
The Woodland Cemetery
Casa Malaparte
National and University Library of Slovenia
Dymaxion House
Golconde, Pondicherry
Baker House
Barragán House and Studio
Turin Exhibition Hall
The Eames House (Case Study House No. 8)
Architect's Second House, São Paulo
Farnsworth House
Säynätsalo Town Hall
L'Unité d'Habitation
Iglesia de la Medalla Milagrosa
Chapelle Notre-Dame du Haut
MIT Chapel
Villa Sarabhai
Sydney Opera House
Palazzetto dello Sport
Seagram Building
Louisiana Museum of Modern Art
National Congress, Brazil
Solomon R. Guggenheim Museum
Los Manantiales
Bacardi Bottling Plant
Couvent Sainte-Marie de La Tourette
TWA Flight Center
Berliner Philharmonie
National Olympic Gymnasium, Tokyo
Salk Institute
The Assembly, Chandigarh
Museu de Arte de São Paulo
Nakagin Capsule Tower
Phillips Exeter Academy Library
Kimbell Art Museum
Castelvecchio Museum
Gallaratese II Apartments
Centre Pompidou
Gehry House
Hongkong and Shanghai Bank Headquarters
Ricola Storage Building
Institut du Monde Arabe
Parc de la Villette
Congrexpo
Guggenheim Museum Bilbao
Piscinas de Marés

拉斯·勒普
（Lars Lerup）

Villa Snellman
Stockholm Public Library
The Woodland Cemetery
Göteborg Town Hall
Stockholm Exhibition of 1930
Chapel of the Resurrection
Court of the National Social Insurance Building
St. Mark's Church in Björkhagen
Church of St. Peter, Klippan
Stockholm Concert Hall
Flower Kiosk, Malmö Cemetery
Vårby Gårds Kyrka
Kulturhuset, Stockholm
The Riksbank Building, Stockholm
Villa Klockberga
Filmhuset, Stockholm
Lunds Konsthall
Malmö Konsthall
Göteborgs Konstmuseum
Svappavaara Housing
Byker Wall
The Ark, London
Brittgården Housing Estate
Allhuset
Villa Erskine
School Churwalden
House Räth
Shelters for Roman Archaeological Site
Saint Benedict Chapel, Sumvitg
Bündner Kunstmuseum, Chur
Gugalun House
Therme Vals
Kunsthaus Bregenz
Bruder Klaus Field Chapel
Serpentine Gallery Pavilion, 2011
Synthes Headquarters, Solothurn
Museum La Congiunta
Novartis Campus, 6 Fabrikstrasse
Swiss Re Tűfihaus
Gartenbad und Schule, Wohlen
Eliot Hall, Washington University
Palestra, Losone
Casa Rezzonico
Edificio Postale, Locarno
Casa Aurora, Lugano
House of Three Women, Beinwil am See
Augustinian Monte Carasso Monastery
Palestra, Monte Carasso
Casa Guidotti
Casa Grossi
Casa Bianchi
Goetz Collection
Signal Box Auf dem Wolf
Dominus Winery
Tate Modern
Laban Dance Centre, London
Forum Building
166 Cottbus Library
De Young
Allianz Arena
Beijing National Stadium
CaixaForum Madrid
1111 Lincoln Road
Museum der Kulturen
Parrish Art Museum
Congrexpo
Netherlands Dance Theatre, Hague
Villa dall'Ava
Nexus World Housing
Kunsthal, Rotterdam
Maison à Bordeaux
Netherlands Embassy, Berlin
Prada Epicenters
Seattle Central Library
Casa da Música
Shenzhen Stock Exchange
Dee and Charles Wyly Theatre
Milstein Hall
CCTV Headquarters
New Court Rothschild Bank
Villa Savoye
Münster City Library
Luxor Theater, Rotterdam
Jewish Museum Berlin
Berlin Memorial to the Murdered Jews of Europe
1234 Howard Street
Architecture and Art Building, Prairie View A&M U.
Guggenheim Museum Bilbao
San Francisco Federal Building
Casa del Fascio
Pabellón de España, Zaragoza
Municipal Auditorium of Teulada
Murcia Town Hall
Viviendas en la Calle Prior
Gimansio del Colegio Maravillas
Casa Ugalde
Peabody Terrace
Igualada Cemetery
BMW Welt
Samitaur Tower

丹尼尔·里伯斯金 / 里伯斯金工作室
（Daniel Libeskind / Studio Libeskind）

Church of the Three Crosses
Säynätsalo Town Hall
Villa Mairea
Paimio Sanatorium
Church of the Light
Barragán House and Studio
AEG Turbine Factory
La Maison de Verre
Le Palais ideal
Het Schip
Steinhaus, Lake Ossiach
Simon Fraser University
Norwegian Glacier Museum
Villa Kise
Hedmark Cathedral Museum
Hongkong and Shanghai Bank Headquarters
Montreal Biosphère
Casa alle Zattere, Venice
Casa Batlló
Casa Milà
Park Güell
Sagrada Família
Guggenheim Museum Bilbao
Hanselmann House
E-1027
Bauhaus Dessau
Gut Garkau Farm
Kreuzberg Tower
Palais Stoclet
Chilehaus
Maison & Atelier Horta
The Woodland Cemetery
SAS Royal Hotel, Copenhagen
National Assembly Building, Bangladesh
Kimbell Art Museum
Salk Institute
Chapelle Notre-Dame du Haut
The Assembly, Chandigarh
Centre Le Corbusier (Heidi Weber House)
Couvent Sainte-Marie de La Tourette
Villa Stein
Villa Savoye
Männistö Church, Kuopio
Church of St. Peter, Klippan
Casa Malaparte
Villa Müller
Glasgow School of Art
Israel Museum
High Museum of Art
Melnikov House
Einstein Tower
Schaubühne
Chiesa dell'Autostrada del Sole
Corso Italia Complex, Milan
Palazzetto dello Sport
Lovell Health House
National Congress, Brazil
Maison Drusch
Bank of China Tower, Hong Kong
Église Notre-Dame du Raincy
Centre Pompidou
Sulphuric Acid Factory, Lubon
Großes Schauspielhaus
Maison Tropicale
Toronto City Hall
Schröderhuis
Fagnano Olona Elementary School
San Cataldo Cemetery
Rudolph Hall
Gateway Arch
TWA Flight Center
Helsinki Central Railway Station
Habitat '67
Museo Canova
Berlin State Library
Sears Tower
Goetheanum
Neue Staatsgalerie
Leicester University Engineering Building
Braun AG Headquarters, Melsungen
Carson, Pirie, Scott and Company Building
Krause Music Store
National Olympic Gymnasium, Tokyo
Hufeisensiedlung
Casa del Fascio
Sydney Opera House
Farnsworth House
New National Gallery
Seagram Building
Tugendhat House
Vanna Venturi House
Mostorg Department Store
Palace of Labor
Kaiser Pavilion, Stadtbahn Station
Edgar J. Kaufmann House (Fallingwater)
Solomon R. Guggenheim Museum
S.C. Johnson & Son Headquarters
Frederick C. Robie House
Unity Temple
Philips Pavilion

格雷格·林恩
（Greg Lynn）

Viipuri Municipal Library
House of the Century
The Woodland Cemetery
Futurama, 1939 New York World's Fair
Maison Bloc
Casa de Vidro
Museu de Arte de São Paulo
Alcuin Library
The Beinecke Rare Book & Manuscript Library
La Maison de Verre
Pacific Design Center
Steinhaus, Lake Ossiach
IBM Pavillion, NY World's Fair
The Eames House (Case Study House No. 8)
Design, Architecture, Art & Planning Building
Art Center College of Design, Pasadena
Dymaxion House
Casa Batlló
Vitra Design Museum
Hanselmann House
Bal Harbor Shops
Gwathmey Residence
Phaeno Science Center
Wall House II
Elbphilharmonie
Palais Stoclet
Retti Candle Shop
Hôtel Solvay
Villa Girasole
Expo '70 Osaka, Festival Plaza
Sendai Mediatheque
Langham Place, Hong Kong
Glass House
Performing Arts Theater, Fort Wayne
Endless House
Seattle Central Library
Maison à Bordeaux
Nakagin Capsule Tower
Bob and Dolores Hope Estate
Carpenter Center for the Visual Arts
Couvent Sainte-Marie de La Tourette
Villa Savoye
Casa Malaparte
Regional Council, Trento
Jewish Museum Berlin
Villa Müller
London Zoo Penguin Pool
Castle Drogo
Hill House, Helensburgh
Splitting, 1974
Fiat Works
Diamond Ranch High School
Rusakov Workers' Club
Petersdorff Department Store
Teatro Regio
Watergate Complex
Hayden Tract
US Air Force Academy Cadet Chapel
Kaufmann House
Das Canoas House
Gustavo Capanema Palace
Fondation Cartier pour l'Art Contemporain
Secession Building
Weissenhof Estate, Haus 5–9
Église Sainte-Bernadette du Banlay
Centre Pompidou
Westin Bonaventure Hotel
Snowdon Aviary
Rooftop Remodeling Falkestrasse
BMW Welt
Maison Tropicale
Schröderhuis
Knights of Columbus Building
Rogers House
Gallaratese II Apartments
Rudolph Residence, 23 Beekman Place
TWA Flight Center
First Christian Church, Columbus
Habitat '67
Berlin State Library
Schindler House
MLC Centre
Shukhov Tower
House of the Future, 1956
Solimene Ceramics Factory
Leicester University Engineering Building
Futuro House
Asilo Sant'Elia
Parc de la Villette
Bagsværd Church
Farnsworth House
Seagram Building
Tugendhat House
Gordon Wu Hall
Postsparkasse
Norris Dam
Indeterminate Façade Building
Solomon R. Guggenheim Museum
S.C. Johnson & Son Headquarters
George Sturges House

维尼·马斯 + 雅各布·范里斯 + 娜塔莉·德弗里斯 /MVRDV 事务所
（Winy Maas + Jacob van Rijs + Nathalie de Vries）

Glass Pavilion
Helsinki Central Railway Station
Großes Schauspielhaus
Einstein Tower
Imperial Hotel, Tokyo
Liverpool Cathedral
Schröderhuis
Zuev Workers' Club
Chrysler Building
Milano Centrale Railway Station
Reichsparteitagsgelände
Battersea Power Station
Molitor Building
New Reich Chancellery
Barragán House and Studio
Glass House
Skylon, Festival of Britain
Royal Festival Hall
Bankside Power Station
Groothandelsgebouw
Hiroshima Peace Memorial Museum
Taliesin West
Unité d'Habitation, Berlin
Maison Louis Carré
Esherick House
Michael Faraday Memorial
National Art Schools of Cuba
La Rinascente, Milan
Nordic Pavillion, Venice
New Saint Michael's Cathedral, Coventry
Carpenter Center for the Visual Arts
Vanna Venturi House
BT Tower
Marina City, Chicago
St. Peter's Seminary, Cardross
History Faculty Building, Cambridge
Orange County Government Centre, Goshen
Kafka Castle
Bensberg Town Hall
Toronto-Dominion Centre
Diagoon Housing
Pimlico Academy
Embassy of the United Kingdom, Rome
Olympic Stadium, Munich
Corviale
Trellick Tower
Brunswick Centre
Robin Hood Gardens
Douglas House
InterAction Centre, Kentish Town
Robarts Library
Sears Tower
Bank of Georgia headquarters, Tbilisi
Jewelry Store Schullin
Hirshhorn Museum
Retti Candle Shop
Yale Center for British Art
Cube Houses, Rotterdam
Prentice Women's Hospital Building
1 United Nations Plaza
Barbican Centre and Estate
CN Tower
Muziekcentrum Vredenburg
Crystal Cathedral
AT&T Building, New York
San Cataldo Cemetery
Bolwoningen
Venetian City Garden
Brixton Recreation Centre
Hongkong and Shanghai Bank Headquarters
Hundertwasserhaus
Tower of Winds, Yokohama
Nemausus
Central Research Institute of Robotics & Cybernetics
French Ministry for the Economy and Finance
Church on the Water
Pyramide du Louvre
Teatro Carlo Felice
Nexus World Housing
Allen Lambert Galleria
Fondation Cartier pour l'Art Contemporain
Paper Church, Kobe
The Hague City Hall
Maggie's Edinburgh
Dancing House
WoZoCo
Getty Center
Blades Residence
British Library
Castalia, The Hague
Maison à Bordeaux
O-Museum, Nagano
Zurichtoren
Borneo 12
Millennium Dome
Kunsthaus Graz
Eden Project
London Millennium Footbridge
Tate Modern
Sagrada Família

槇文彦
(Fumihiko Maki)

Beurs van Berlage
Glasgow School of Art
Postsparkasse
Frederick C. Robie House
AEG Turbine Factory
Gamble House
Casa Milà
Michaelerplatz House (Looshaus)
Einstein Tower
Schindler House
Schröderhuis
Lovell Beach House
Bauhaus Dessau
Stockholm Public Library
Villa Moller
German Pavilion (Barcelona Pavilion)
Van Nelle Factory
Villa Stein
Tugendhat House
Stockholm Exhibition of 1930
Villa Müller
Villa Savoye
La Maison de Verre
Casa del Fascio
Edgar J. Kaufmann House (Fallingwater)
S.C. Johnson & Son Headquarters
Villa Mairea
The Woodland Cemetery
Dymaxion House
L'Unité d'Habitation
The Eames House (Case Study House No. 8)
Farnsworth House
Baker House
Barragán House and Studio
Turin Exhibition Hall
Glass House
Crown Hall
Säynätsalo Town Hall
Lever House
The Assembly, Chandigarh
Muuratsalo Experimental House
Chapelle Notre-Dame du Haut
St. Mark's Church in Björkhagen
Municipal Orphanage
Sydney Opera House
Mill Owners' Association Building
Berliner Philharmonie
Couvent Sainte-Marie de La Tourette
Seagram Building
Museu de Arte de São Paulo
National Congress, Brazil
Solomon R. Guggenheim Museum
Leicester University Engineering Building
Salk Institute
Stahl House (Case Study House No. 22)
Halen Estate
TWA Flight Center
US Air Force Academy Cadet Chapel
Carpenter Center for the Visual Arts
Vanna Venturi House
John Deere World Headquarters
Querini Stampalia Renovation
The Beinecke Rare Book & Manuscript Library
Ford Foundation Headquarters
Free University of Berlin
National Olympic Gymnasium, Tokyo
Peabody Terrace
Sea Ranch Condominium Complex
Centre Le Corbusier (Heidi Weber House)
The Breuer Building (Whitney Museum)
Piscinas de Marés
Habitat '67
Hillside Terrace Complex I–VI
Smith House
New National Gallery
Olympic Stadium, Munich
Brion Family Tomb
House III (Miller House)
Nakagin Capsule Tower
Museum of Modern Art, Gunma
Centre Pompidou
Kimbell Art Museum
Phillips Exeter Academy Library
Centraal Beheer Building
Bagsværd Church
World Trade Center Towers
SESC Pompéia
Teatro del Mondo
Gehry House
Hongkong and Shanghai Bank Headquarters
Lloyd's of London
Parc de la Villette
National Assembly Building, Bangladesh
Neue Staatsgalerie
National Museum of Roman Art
The Menil Collection
Institut du Monde Arabe
Rooftop Remodeling Falkestrasse
Guggenheim Museum Bilbao
Sendai Mediatheque

米夏埃尔·马尔灿
(Michael Maltzan)

Leicester University Engineering Building
Beurs van Berlage
National Farmers Bank
AEG Turbine Factory
Posen Tower
Postsparkasse
Fagus Factory
Park Güell
Fiat Works
Schindler House
Ozenfant House
Schröderhuis
Schocken Department Store, Stuttgart
Bauhaus Dessau
Stockholm Public Library
Karl-Marx-Hof
Villa Moller
Melnikov House
German Pavilion (Barcelona Pavilion)
Open-air School, Amsterdam
Villa Savoye
Boots Pharmaceutical Factory
London Zoo Penguin Pool
Viipuri Municipal Library
Casa del Fascio
S.C. Johnson & Son Headquarters
W.E. Oliver House
Casa Malaparte
The Woodland Cemetery
Casa del Puente
Baker House
Barragán House and Studio
Glass House
Lever House
L'Unité d'Habitation
Muuratsalo Experimental House
Yale University Art Gallery
General Motors Technical Center
Villa Planchart
Gávea Housing Complex
Torre Velasca
860–880 Lake Shore Drive
Inland Steel Building
National Congress, Brazil
Couvent Sainte-Marie de La Tourette
The Beinecke Rare Book & Manuscript Library
Berliner Philharmonie
Vanna Venturi House
Peabody Terrace
National Olympic Gymnasium, Tokyo
National Museum of Anthropology
Sea Ranch Condominium Complex
The Economist Building, London
Gwathmey Residence
Piscinas de Marés
The Breuer Building (Whitney Museum)
Habitat '67
Hanselmann House
New National Gallery
Ford Foundation Headquarters
Art Center College of Design, Pasadena
Andrew Melville Hall
Mummers Theater
House II (Vermont House)
Nakagin Capsule Tower
Government Service Center, Boston
Robin Hood Gardens
Phillips Exeter Academy Library
Center for the Arts, Wesleyan University
Kresge College
Free University of Berlin
Gallaratese II Apartments
Centraal Beheer Building
Whig Hall, Princeton University
Willis Faber and Dumas Headquarters
Uehara House
John Hancock Tower
SESC Pompéia
Neue Staatsgalerie
Loyola Law School Campus
National Museum of Roman Art
Capela São Pedro Apóstolo
The Menil Collection
Institut du Monde Arabe
Museum of Sculpture, São Paulo
Vitra Design Museum Void
Space/Hinged Space Housing
Santa Maria Church de Canaveses
Nexus World Housing
Santa Caterina Market
Fondation Cartier pour l'Art Contemporain
Hayden Tract
Educatorium
21st Century Museum, Kanazawa
Nelson-Atkins Museum of Art Addition
Tama Art University Library
Iberê Camargo Foundation
Teshima Art Museum
1111 Lincoln Road
Madrid Public Housing

汤姆·梅恩 / 墨菲西斯事务所
(Thom Mayne / Morphosis)

Postsparkasse
Schindler House
Fiat Works
Schröderhuis
Cité Frugès Housing Complex
Karl-Marx-Hof
Lovell Health House
Stockholm Public Library
Villa Moller
German Pavilion (Barcelona Pavilion)
Villa Savoye
La Maison de Verre
Villa Girasole
Edgar J. Kaufmann House (Fallingwater)
Casa del Fascio
The Woodland Cemetery
Casa Malaparte
Dymaxion House
Turin Exhibition Hall
Glass House
The Eames House (Case Study House No. 8)
Baker House
Muuratsalo Experimental House
Farnsworth House
L'Unité d'Habitation
Lever House
Pedregulho Housing Complex
Crown Hall
Maisons Jaoul
Chapelle Notre-Dame du Haut
Seagram Building
National Congress, Brazil
Philips Pavilion
Solomon R. Guggenheim Museum
Cité d'Habitacion of Carrières Centrales
Stahl House (Case Study House No. 22)
Couvent Sainte-Marie de La Tourette
Municipal Orphanage
Halen Estate
TWA Flight Center
Querini Stampalia Renovation
Berliner Philharmonie
Vanna Venturi House
National Olympic Gymnasium, Tokyo
Peabody Terrace
Rudolph Hall
Salk Institute
The Assembly, Chandigarh
Centre Le Corbusier (Heidi Weber House)
Piscinas de Marés
Bank of London and South America
Habitat '67
Smith House
Museu de Arte de São Paulo
Ford Foundation Headquarters
Kowloon Walled City
History Faculty Building, Cambridge
Benacerraf House Project
Flower Kiosk, Malmö Cemetery
Petrobras Headquarters
The Beinecke Rare Book & Manuscript Library
House III (Miller House)
Florey Building
Nakagin Capsule Tower
Olympic Stadium, Munich
Phillips Exeter Academy Library
Kimbell Art Museum
Free University of Berlin
Byker Wall
Centraal Beheer Building
Whig Hall, Princeton University
Bagsværd Church
Brion Family Tomb
Centre Pompidou
SESC Pompéia
Hedmark Cathedral Museum
Gehry House
Teatro del Mondo
Atheneum, New Harmony
Koshino House
Vietnam Veterans Memorial
National Assembly Building, Bangladesh
Neue Staatsgalerie
Lloyd's of London
Hongkong and Shanghai Bank Headquarters
Steinhaus, Lake Ossiach
Rooftop Remodeling Falkestrasse
National Museum of Roman Art
Void Space/Hinged Space Housing
Igualada Cemetery
Parc de la Villette
Maison à Bordeaux
Samitaur Tower
Guggenheim Museum Bilbao
UFA Cinema Centre
Jewish Museum Berlin
Bordeaux Law Courts
Diamond Ranch High School
Yokohama International Port Terminal
Sendai Mediatheque

理查德·迈耶
(Richard Meier)

Sydney Opera House
Maison Louis Carré
Hongkong and Shanghai Bank Headquarters
Church of the Light
Magney House
Tate Modern
Centre Pompidou
Lloyd's of London
Barragán House and Studio
Camino Real Hotel, Mexico City
Nara Centennial Hall
Kaze-no-Oka Crematorium
Slither Housing
Petronas Towers
Kiasma
Wall House II
Serralves Museum of Contemporary Art
National Gallery of Art, East Building
BMW Central Building
De Blas House
Neuer Zollhof
Jewish Museum Berlin
Alfonse M. D'Amato United States Courthouse
Neugebauer House
National Farmers Bank
Chrysler Building
The Woodland Cemetery
Bauhaus Dessau
AEG Turbine Factory
Beurs van Berlage
Open-air School, Amsterdam
Van Nelle Factory
Flatiron Building
La Maison de Verre
Museum of the Great War
Zuev Workers' Club
Kanchanjunga Apartments
Gustavo Capanema Palace
Van Doesburg House
Palau de la Música Catalana
Hilversum Town Hall
The Eames House (Case Study House No. 8)
Berlin Memorial to the Murdered Jews of Europe
Municipal Orphanage
Figini House
Josef Frank House, Vienna
Casa Milà
Guggenheim Museum Bilbao
Woolworth Building
E-1027
Gamble House
Métropolitain Entrance
Palais Stoclet
Chapel of St. Ignatius
Glass House
Seagram Building
Rockefeller Center
Villa Stein
Chapelle Notre-Dame du Haut
Salk Institute
Villa dall'Ava
East Bohemian Museum
Casa Malaparte
Villa Müller
Castle Drogo
Glasgow School of Art
Villa Allatini and Villa de Daniel Dreyfuss
Bingham House, Santa Barbara
Pennsylvania Station
Getty Center
Melnikov House
Einstein Tower
German Pavilion (Barcelona Pavilion)
National Museum of Roman Art
Lovell Health House
Palácio do Planalto
Joseph Maria Olbrich's House
Weissenhof Estate, Haus 5-9
Church of the Most Sacred Heart of Our Lord
Großes Schauspielhaus
Schröderhuis
Ford Foundation Headquarters
Rudolph Hall
TWA Flight Center
Cranbrook Academy of Art
Castelvecchio Museum
Berliner Philharmonie
Lovell Beach House
Fundació Joan Miró
Lever House
The Economist Building, London
Leicester University Engineering Building
Glass Pavilion
Casa del Fascio
Postsparkasse
Unity Temple
Frederick C. Robie House
Edgar J. Kaufmann House (Fallingwater)
S.C. Johnson & Son Headquarters
Solomon R. Guggenheim Museum

拉斐尔·莫内欧
(Rafael Moneo)

Secession Building
Rue Franklin Apartments
Glasgow School of Art
Gamble House
AEG Turbine Factory
Frederick C. Robie House
Michaelerplatz House (Looshaus)
Posen Tower
Palais Stoclet
Postsparkasse
Festival Hall, Hellerau
Spaardammerplantsoen
Rashtrapati Bhavan
Schindler House
Ca' Brutta
Casa Milà
Schröderhuis
Bauhaus Dessau
Villa Müller
Lovell Health House
Weiße Stadt, Berlin
Villa Savoye
Van Nelle Factory
ADGB Trade Union School
Schocken Department Store, Stuttgart
Boots Pharmaceutical Factory
Hilversum Town Hall
Casa Malaparte
Tugendhat House
Casa de las Flores, Madrid
Edificio Carrión
Cineac
Casa del Fascio
Villa Mairea
Rockefeller Center
The Woodland Cemetery
Göteborg Town Hall
S.C. Johnson & Son Headquarters
National and University Library of Slovenia
Resurrection Chapel, Turku
Casa del Puente
XYZ Buildings, New York
Lever House
Borsalino Housing
Capilla de las Capuchinas
Pedregulho Housing Complex
Casa Il Girasole
Mill Owners' Association Building
Munkegaard School
Civil Government Building, Tarragona
Torre Velasca
Sydney Opera House
Seagram Building
Municipal Orphanage
Yale University Art Gallery
The Breuer Building (Whitney Museum)
Casa Ugalde
National Congress, Brazil
Kline Biology Tower
Rudolph Hall
Retti Candle Shop
TWA Flight Center
La Rinascente, Milan
Leicester University Engineering Building
Berliner Philharmonie
Querini Stampalia Renovation
Vanna Venturi House
Peabody Terrace
National Olympic Gymnasium, Tokyo
The Economist Building, London
Sea Ranch Condominium Complex
Torre Blancas
Museu de Arte de São Paulo
Ford Foundation Headquarters
John Hancock Tower
Indian Institute of Management, Ahmedabad
House II (Vermont House)
Centraal Beheer Building
Twin Parks Northeast Housing
Tokyo Metropolitan Art Museum
Gallaratese II Apartments
Centre Pompidou
Sainsbury Center for Visual Arts
Hedmark Cathedral Museum
Faculty of Architecture of the University of Porto
Portland Building
Neue Staatsgalerie
Loyola Law School Campus
Church of the Light
Bank of China Tower, Hong Kong
Igualada Cemetery
Private House, Corrubedo
Portuguese National Pavilion
Maison à Bordeaux
Therme Vals
Dominus Winery
Guggenheim Museum Bilbao
Kiasma
Gifu Kitagata Apartment Building
Sendai Mediatheque

埃里克·欧文·莫斯
(Eric Owen Moss)

Larkin Company Administration Building
AEG Turbine Factory
Glasgow School of Art
Park Güell
Großes Schauspielhaus
Einstein Tower
Fiat Works
Shukhov Tower
Schindler House
Hutfabrik Friedrich Steinberg, Herrmann & Co.
Imperial Hotel, Tokyo
Schröderhuis
Magazzini Generali, Chiasso
Lovell Beach House
Bauhaus Dessau
Gut Garkau Farm
Weissenhof Estate, Exhibition of 1927
Villa Moller
Rusakov Workers' Club
Narkomfin Building
Melnikov House
Lovell Health House
German Pavilion (Barcelona Pavilion)
Villa Savoye
Ogonyok Magazine Printing Plant
Edgar J. Kaufmann House (Fallingwater)
Casa del Fascio
S.C. Johnson & Son Headquarters
Aarhus City Hall
The Eames House (Case Study House No. 8)
Glass House
Farnsworth House
Säynätsalo Town Hall
Lever House
Capilla de las Capuchinas
Watts Towers
Iglesia de la Medalla Milagrosa
Chapelle Notre-Dame du Haut
Sydney Opera House
National Congress, Brazil
National Library, Brasília
North Shore Beach and Yacht Club
Church of the Three Crosses
Municipal Orphanage
Couvent Sainte-Marie de La Tourette
Stahl House (Case Study House No. 22)
St. John's Abbey, Collegeville
TWA Flight Center
Querini Stampalia Renovation
John Deere World Headquarters
Berliner Philharmonie
Leicester University Engineering Building
Palm Springs Aerial Tramway Valley Station
Sheats Goldstein Residence
Carpenter Center for the Visual Arts
Peabody Terrace
Rudolph Hall
Vanna Venturi House
National Olympic Gymnasium, Tokyo
Frey House II
The Assembly, Chandigarh
Salk Institute
National Art Schools of Cuba
Retti Candle Shop
Shizuoka Press and Broadcasting Center
Habitat '67
Knights of Columbus Building
Benacerraf House Project
Geisel Library
Expo' 70 Tower, Osaka
Phillips Exeter Academy Library
Florey Building
Aula Paolo VI
Nakagin Capsule Tower
Olympic Stadium, Munich
Free University of Berlin
San Francisco State Student Center
Douglas House
Museum of Modern Art, Gunma
Gallaratese II Apartments
Kitakyushu International Conference Center
Indian Institute of Management, Ahmedabad
Bagsværd Church
Transvaal House
Gehry House
Ramot Polin
Steinhaus, Lake Ossiach
National Museum of Roman Art
Marina City, Chicago
House of the Suicide
Rooftop Remodeling Falkestrasse
Wexner Center for the Visual Arts
Igualada Cemetery
Berliner Philharmonie
Diamond Ranch High School
Chapel of St. Ignatius
Jean-Marie Tjibaou Cultural Centre
Stealth, Culver City
Sagrada Família
MIT Chapel

恩里克·诺尔滕/TEN 事务所
(Enrique Norten / TEN Arquitectos)

Rue Franklin Apartments
Larkin Company Administration Building
AEG Turbine Factory
Glasgow School of Art
Gamble House
Frederick C. Robie House
Casa Milà
Fagus Factory
Park Güell
Michaelerplatz House (Looshaus)
Einstein Tower
Ozenfant House
Soviet Pavilion
Lovell Beach House
Sagrada Família
Bauhaus Dessau
Stockholm Public Library
German Pavilion (Barcelona Pavilion)
Rusakov Workers' Club
Narkomfin Building
Casa Estudio Diego Rivera y Frida Kahlo
Villa Müller
Villa Savoye
Edgar J. Kaufmann House (Fallingwater)
Casa del Fascio
Asilo Sant'Elia
S.C. Johnson & Son Headquarters
Villa Mairea
Súper Servicio Lomas
Centro Urbano Presidente Alemán
Dymaxion House
New Gourna Village Mosque
Barragán House and Studio
Turin Exhibition Hall
The Eames House (Case Study House No. 8)
Museo Experimental El Eco
Biblioteca Central, UNAM
Rectoría Building, UNAM
Olympic University Stadium, UNAM
Faculty of Philosophy and Letters, UNAM
Gante Indoor Garage
Torre Latinoamericana
Conjunto Aristos
Torres de Satélite
Farnsworth House
Lever House
Muuratsalo Experimental House
Iglesia de la Medalla Milagrosa
Chapelle Notre-Dame du Haut
The Assembly, Chandigarh
MIT Chapel
Sydney Opera House
Mill Owners' Association Building
Palazzetto dello Sport
Cube House, Köln
National Congress, Brazil
Los Manantiales
Conjunto Urbano Nonoalco Tlatelolco
Palacio de los Deportes
Celanese Mexicana
Capilla de las Capuchinas
Couvent Sainte-Marie de La Tourette
Carpenter Center for the Visual Arts
Berliner Philharmonie
Otaniemi Technical University
Vanna Venturi House
Centre Le Corbusier (Heidi Weber House)
Salk Institute
Piscinas de Marés
Edificio Copan
The Breuer Building (Whitney Museum)
Habitat '67
Museu de Arte de São Paulo
New National Gallery
Cuadra San Cristóbal
Art Center College of Design, Pasadena
Itamaraty Palace
Kimbell Art Museum
Castelvecchio Museum
Gallaratese II Apartments
Marie Short House
SESC Pompéia
Centre Pompidou
Teatro del Mondo
Museo Rufino Tamayo
Koshino House
Hongkong and Shanghai Bank Headquarters
National Museum of Roman Art
The Menil Collection
Televisa Mixed Use Building
Escuela Nacional de Teatro
Casa LE
Hotel Habita, Mexico City
Igualada Cemetery
Kunsthal, Rotterdam
Goetz Collection
Therme Vals
Kunsthaus Bregenz
Chapel of St. Ignatius
Guggenheim Museum Bilbao

若泽·乌贝里
(José Oubrerie)

L'Esprit Nouveau Pavilion
Villa La Roche
Villa Savoye
L'Unité d'Habitation
Chapelle Notre-Dame du Haut
Villa Sarabhai
Villa Shodhan
Mill Owners' Association Building
Maisons Jaoul
Couvent Sainte-Marie de La Tourette
The Assembly, Chandigarh
Carpenter Center for the Visual Arts
Centre Le Corbusier (Heidi Weber House)
German Pavilion (Barcelona Pavilion)
Tugendhat House
Weissenhof Estate, Exhibition of 1927
Farnsworth House
Crown Hall
New National Gallery
Villa Mairea
Studio Aalto
Stockholm Public Library
Baker House
Church of the Three Crosses
Unity Temple
Frederick C. Robie House
Edgar J. Kaufmann House (Fallingwater)
S.C. Johnson & Son Headquarters
Millard House
Richards Medical Research Laboratories
Phillips Exeter Academy Library
Kimbell Art Museum
Salk Institute
Indian Institute of Management, Ahmedabad
National Assembly Building, Bangladesh
Villa Müller
Villa Moller
Café Museum
Casa Milà
Park Güell
Gehry House
Vitra Design Museum
Guggenheim Museum Bilbao
Schindler House
Lovell Beach House
Leicester University Engineering Building
Andrew Melville Hall
Querini Stampalia Renovation
Castelvecchio Museum
Brion Family Tomb
Lever House
The Beinecke Rare Book & Manuscript Library
Wexner Center for the Visual Arts
House III (Miller House)
House VI
Kiasma
Chapel of St. Ignatius
Nelson-Atkins Museum of Art Addition
Kate Mantilini, Beverly Hills
Venice III House
Diamond Ranch High School
Villa dall'Ava
Maison à Bordeaux
Congrexpo
Kunsthal, Rotterdam
Fondation Cartier pour l'Art Contemporain
Institut du Monde Arabe
Rooftop Remodeling Falkestrasse
UFA Cinema Centre
Lloyd's of London
Vitra Fire Station
Miller House, Lexington
Free University of Berlin
Centre Pompidou
TWA Flight Center
Municipal Orphanage
Igualada Cemetery
Allied Bank Tower
Rue Franklin Apartments
Glasgow School of Art
Bauhaus Dessau
Narkomfin Building
E-1027
La Maison de Verre
The Eames House (Case Study House No. 8)
Casa Malaparte
Schröderhuis
Einstein Tower
Casa del Fascio
Lovell Health House
Gamble House
Rusakov Workers' Club
Berliner Philharmonie
The Woodland Cemetery
St. Mark's Church in Björkhagen
Barragán House and Studio
National Congress, Brazil
Glass House
Stahl House (Case Study House No. 22)
Goetheanum

西萨·佩里
(César Pelli)

Beurs van Berlage
Rue Franklin Apartments
AEG Turbine Factory
Glasgow School of Art
Gamble House
Union Bank, Columbus, Wisconsin
Larkin Company Administration Building
Unity Temple
Frederick C. Robie House
Fagus Factory
Pennsylvania Station
Het Schip
Schindler House
Fiat Works
Schröderhuis
Cité Frugès Housing Complex
Lovell Beach House
Karl-Marx-Hof
Lovell Health House
German Pavilion (Barcelona Pavilion)
Einstein Tower
Bauhaus Dessau
Goetheanum
Villa Savoye
La Maison de Verre
Edgar J. Kaufmann House (Fallingwater)
Casa del Fascio
S.C. Johnson & Son Headquarters
Chrysler Building
Rockefeller Center
The Woodland Cemetery
Dymaxion House
Baker House
Glass House
The Eames House (Case Study House No. 8)
Cranbrook Academy of Art
Carpenter Center for the Visual Arts
Farnsworth House
L'Unité d'Habitation
Lever House
Chapelle Notre-Dame du Haut
Sydney Opera House
Seagram Building
Solomon R. Guggenheim Museum
Crown Hall
Maisons Jaoul
Mill Owners' Association Building
Miller House, Columbus
Church of Cristo Obrero
Stahl House (Case Study House No. 22)
Couvent Sainte-Marie de La Tourette
TWA Flight Center
Snowdon Aviary
John Deere World Headquarters
Berliner Philharmonie
Leicester University Engineering Building
National Olympic Gymnasium, Tokyo
Rudolph Hall
Vanna Venturi House
Salk Institute
The Assembly, Chandigarh
Bank of London and South America
Smith House
Hyatt Regency, Atlanta
Ford Foundation Headquarters
Museu de Arte de São Paulo
Knights of Columbus Building
Andrew Melville Hall
Fire Station No. 4, Columbus
COMSAT Laboratories
Sea Ranch Condominium Complex
Nakagin Capsule Tower
Olympic Stadium, Munich
Phillips Exeter Academy Library
Snyderman House
Kimbell Art Museum
Free University of Berlin
House VI
Casa Bianchi
Whig Hall, Princeton University
Centre Pompidou
Brion Family Tomb
Gehry House
Teatro del Mondo
Very Large Array
Vietnam Veterans Memorial
National Assembly Building, Bangladesh
Lloyd's of London
Hongkong and Shanghai Bank Headquarters
National Museum of Roman Art
Steinhaus, Lake Ossiach
Igualada Cemetery
Parc de la Villette
Congrexpo
Samitaur Tower
Guggenheim Museum Bilbao
Diamond Ranch High School
Jewish Museum Berlin
Sendai Mediatheque
Petronas Towers

多米尼克·佩罗
(Dominique Perrault)

Métropolitain Entrance
Park Güell
Halle Tony Garnier
AEG Turbine Factory
Glasgow School of Art
Ford Motor Company Glass Manufacturing Plant
Einstein Tower
Église Notre-Dame du Raincy
Fiat Works
Chilehaus
Tristan Tzara House
La Maison de Verre
Rusakov Workers' Club
Stockholm Public Library
Paimio Sanatorium
German Pavilion (Barcelona Pavilion)
Casa Malaparte
Chrysler Building
Gratte-ciel, Villeurbanne
Villa Cavrois
London Zoo Penguin Pool
Gustavo Capanema Palace
Palais d'Iéna
Rockefeller Center
S.C. Johnson & Son Headquarters
Palazzo della Civiltà Italiana
Glass House
Barragán House and Studio
General Motors Technical Center
Farnsworth House
United Nations Headquarters
L'Unité d'Habitation
Annakirche, Düren
St. Mary's Cathedral, Tokyo
Chapelle Notre-Dame du Haut
Maison des Jours Meilleurs
Climat de France, Algiers
Complesso Olivetti
Atomium
Seagram Building
Sky House, Tokyo
Pirelli Tower
Valle de los Caídos
Stahl House (Case Study House No. 22)
Solomon R. Guggenheim Museum
Capilla de las Capuchinas
Habitat '67
Snowdon Aviary
Palazzo del Lavoro
Nordic Pavillion, Venice
Mémorial des Martyrs de la Déportation
The Breuer Building (Whitney Museum)
Salk Institute
Église Sainte-Bernadette du Banlay
Piscinas de Marés
Montreal Biosphère
New National Gallery
John Hancock Center
Marina City, Chicago
Hew Offices, Hamburg
Maison d'Iran
FAU, University of São Paulo
Fondation Maison des sciences de l'homme
United States Pavilion Expo '70, Osaka
Danmarks Nationalbank, Copenhagen
French Communist Party Headquarters, Paris
Olympic Stadium, Munich
Phillips Exeter Academy Library
Nakagin Capsule Tower
Sydney Opera House
White U
Centre Pompidou
Berlin State Library
SESC Pompéia
Hongkong and Shanghai Bank Headquarters
Church on the Water
Lloyd's of London
Church of the Light
Bonjour Tristesse, Berlin
Stone House, Tavole
Pyramide du Louvre
Grande Arche
Fondation Cartier pour l'Art Contemporain
Kunsthal, Rotterdam
Museum of Sculpture, São Paulo
Goetz Collection
Faculty of Architecture of the University of Porto
Sofitel, Tokyo
Velodrome and Olympic Swimming Pool, Berlin
Sendai Mediatheque
National Library of France
Tate Modern
Acros Building
Yokohama International Port Terminal
Guggenheim Museum Bilbao
Therme Vals
Gifu Kitagata Apartment Building
Maison à Bordeaux
21st Century Museum Kanazawa
Jewish Museum Berlin

卡梅·皮诺斯
（Carme Pinós）

Larkin Company Administration Building
AEG Turbine Factory
Glasgow School of Art
Casa Milà
Park Güell
Einstein Tower
Fiat Works
Van Nelle Factory
Schröderhuis
Soviet Pavilion
Lovell Beach House
Villa Moller
E-1027
German Pavilion (Barcelona Pavilion)
Rusakov Workers' Club
Stockholm Exhibition of 1930
Villa Savoye
La Maison de Verre
Edgar J. Kaufmann House (Fallingwater)
S.C. Johnson & Son Headquarters
Arve Bridge
The Woodland Cemetery
Casa Malaparte
Baker House
Turin Exhibition Hall
The Eames House (Case Study House No. 8)
Architect's Second House, São Paulo
Farnsworth House
Säynätsalo Town Hall
L'Unité d'Habitation
Cité d'Habitacion of Carrières Centrales
Chapelle Notre-Dame du Haut
The Assembly, Chandigarh
St. Mark's Church in Björkhagen
MIT Chapel
Sydney Opera House
Richards Medical Research Laboratories
Palazzetto dello Sport
Torre Velasca
Seagram Building
Museum of Modern Art, Rio de Janeiro
Solomon R. Guggenheim Museum
Los Manantiales
Pepsi-Cola Building
Couvent Sainte-Marie de La Tourette
Snowdon Aviary
Carpenter Center for the Visual Arts
TWA Flight Center
Berliner Philharmonie
Leicester University Engineering Building
National Olympic Gymnasium, Tokyo
Otaniemi Technical University
The Economist Building, London
Gwathmey Residence
Salk Institute
High Court, Assembly, Chandigarh
Montessori School, Delft
Sonsbeek Pavilion
The Breuer Building (Whitney Museum)
Norman Fisher House
Torre Blancas
Ford Foundation Headquarters
Museu de Arte de São Paulo
New National Gallery
History Faculty Building, Cambridge
Art Center College of Design, Pasadena
Knights of Columbus Building
Mivtachim Sanitarium
Florey Building
Can Lis, Majorca
Olympic Stadium, Munich
Phillips Exeter Academy Library
Castelvecchio Museum
Free University of Berlin
Douglas House
House VI
Marie Short House
Bagsvaerd Church
Sainsbury Center for Visual Arts
World Trade Center Towers
SESC Pompéia
Centre Pompidou
Brion Family Tomb
Gehry House
Teatro del Mondo
Warehouse, Port of Montevideo
National Assembly Building, Bangladesh
Hongkong and Shanghai Bank Headquarters
Church of the Light
Rooftop Remodeling Falkestrasse
Igualada Cemetery
Storefront for Art and Architecture
Signal Box Auf dem Wolf
Vitra Fire Station
Therme Vals
WoZoCo
Guggenheim Museum Bilbao
Diamond Ranch High School
Yokohama International Port Terminal
American Folk Art Museum

詹姆斯·S·波尔舍克
（James S. Polshek）

Rose Center for Earth and Space
Clinton Presidential Center
Larkin Company Administration Building
Glasgow School of Art
Unity Temple
Casa Milà
Imperial Hotel, Tokyo
Schindler House
Van Nelle Factory
Schröderhuis
Café De Unie
Sagrada Família
Villa Stein
German Pavilion (Barcelona Pavilion)
Villa Müller
Tugendhat House
Villa Savoye
Hilversum Town Hall
Paimio Sanatorium
La Maison de Verre
Edgar J. Kaufmann House (Fallingwater)
S.C. Johnson & Son Headquarters
Villa Mairea
Aarhus City Hall
Barragán House and Studio
The Eames House (Case Study House No. 8)
Farnsworth House
L'Unité d'Habitation
Chapelle Notre-Dame du Haut
Villa Sarabhai
Sydney Opera House
Maisons Jaoul
Mill Owners' Association Building
Richards Medical Research Laboratories
Philips Pavilion
National Library, Brasília
Solomon R. Guggenheim Museum
Taliesin West
Torres de Satélite
Maison Louis Carré
Pepsi-Cola Building
Couvent Sainte-Marie de La Tourette
Municipal Orphanage, Amsterdam
Halen Estate
Nordic Pavilion, Venice
TWA Flight Center
Querini Stampalia Renovation
Gateway Arch
John Deere World Headquarters
Berliner Philharmonie
Leicester University Engineering Building
Otaniemi Technical University
Rudolph Hall
National Art Schools of Cuba
Salk Institute
The Assembly, Chandigarh
Montessori School, Delft
Retti Candle Shop
Ford Foundation Headquarters
History Faculty Building, Cambridge
National Assembly Building, Bangladesh
Indian Institute of Management, Ahmedabad
Teijin Institute for Biomedical Research
Charles De Gaulle Airport, Terminal 1 & 2A–2F
Can Lis, Majorca
Phillips Exeter Academy Library
Franklin D. Roosevelt Four Freedoms Park
Kimbell Art Museum
Castelvecchio Museum
Museum of Modern Art, Gunma
Yale Center for British Art
Town Hall, Logroño
Willis Faber and Dumas Headquarters
Centre Pompidou
Vietnam Veterans Memorial
Smith House
Lloyd's of London
Hongkong and Shanghai Bank Headquarters
The Menil Collection
Institut du Monde Arabe
Pyramide du Louvre
TEPIA Science Pavilion, Tokyo
Neue Staatsgalerie
Norwegian Glacier Museum
Fondation Cartier pour l'Art Contemporain
Neurosciences Institute, La Jolla
Therme Vals
Chapel of St. Ignatius
Jean-Marie Tjibaou Cultural Centre
Diamond Ranch High School
O-Museum, Nagano
Parco della Musica
Mercedes-Benz Museum
Tama Art University Library
CaixaForum Madrid
41 Cooper Square
Parc de la Villette
Bordeaux Law Courts
American Folk Art Museum
Kiasma

莫妮卡·庞塞·德莱昂
（Mónica Ponce de León）

Park Güell
Larkin Company Administration Building
AEG Turbine Factory
Broomfield Rowhouse
Casa Milà
Widener Memorial Library
Ford Motor Company Glass Manufacturing Plant
The Woodland Cemetery
E-1027
Schröderhuis
Bauhaus Dessau
Boston Avenue Methodist Church
Weissenhof Estate, Haus 5-9
Villa Moller
German Pavilion (Barcelona Pavilion)
Lovell Health House
Open-Air School, Amsterdam
Villa Savoye
Casa del Fascio
S.C. Johnson & Son Headquarters
Villa Mairea
Casa del Puente
The Eames House (Case Study House No. 8)
General Motors Technical Center
Lever House
Casa Il Girasole
Casa de Vidro
Pedregulho Housing Complex
L'Unité d'Habitation
Aula Magna, Caracas
Edificio Copan
The Economist Building, London
Eduardo Guinle Park Housing
Seagram Building
Chapelle Notre-Dame du Haut
St. Mark's Church in Björkhagen
Glass House
Pryor House
Villa Planchart
Los Manantiales
Bank of London and South America
Solomon R. Guggenheim Museum
Church of Cristo Obrero
Halen Estate
Ichinomiya Rowhouses
National Art Schools of Cuba
Theme Building
Aalto-Hochhaus, Bremen
TWA Flight Center
Nordic Pavillion, Venice
Rudolph Hall
Donald Pollock House
Berliner Philharmonie
Marina City, Chicago
Chiesa dell'Autostrada del Sole
University Reformed Church, Ann Arbor
Peabody Terrace
Vanna Venturi House
Phillips Exeter Academy Library
John Hancock Center
Saint Francis de Sales Church
Centraal Beheer Building
John Hancock Tower
Oakland Museum of California
World Trade Center Towers
University Art Museum, Berkeley
Sears Tower
San Cataldo Cemetery
Centre Pompidou
Nakagin Capsule Tower
Kimbell Art Museum
Sydney Opera House
Embassy of the United States, Tokyo
Casa Gilardi
Allen Memorial Art Museum, Oberlin
Thorncrown Chapel
Musée d'Orsay
National Museum of Roman Art
San Juan Capistrano Library
Hongkong and Shanghai Bank Headquarters
Parc de la Villette
Igualada Cemetery
Ricola Storage Building
Housing Schilderswijk West
Tokyo Metropolitan Gymnasium
Nexus World Housing
Void Space/Hinged Space Housing
Robert F. Wagner, Jr. Park, New York
Villa VPRO
Vitra Fire Station
Yokohama International Port Terminal
Bruton Barr Library
Santa Caterina Market
Walt Disney Concert Hall
Nomentana Residence
Diamond Ranch High School
N Museum, Wakayama
Jewish Museum Berlin
Hayden Tract
City of Culture of Galicia

安托万·普雷多克
（Antoine Predock）

Larkin Company Administration Building
AEG Turbine Factory
Glasgow School of Art
Gamble House
Unity Temple
Frederick C. Robie House
Casa Milà
Einstein Tower
Schindler House
Fiat Works
Sagrada Família
E-1027
Goetheanum
German Pavilion (Barcelona Pavilion)
Zarzuela Hippodrome
Tugendhat House
Villa Savoye
Hilversum Town Hall
La Maison de Verre
Villa Girasole
Edgar J. Kaufmann House (Fallingwater)
S.C. Johnson & Son Headquarters
Villa Mairea
Arve Bridge
The Woodland Cemetery
Casa Malaparte
Dymaxion House
New Gourna Village Mosque
Barragán House and Studio
Turin Exhibition Hall
The Eames House (Case Study House No. 8)
Farnsworth House
Säynätsalo Town Hall
Chapelle Notre-Dame du Haut
The Assembly, Chandigarh
MIT Chapel
Sydney Opera House
Seagram Building
Louisiana Museum of Modern Art
National Congress, Brazil
Solomon R. Guggenheim Museum
Hiroshima Peace Memorial Museum
Los Manantiales
Pepsi-Cola Building
Stahl House (Case Study House No. 22)
Couvent Sainte-Marie de La Tourette
Prairie Chicken House
Halen Estate
US Air Force Academy Cadet Chapel
TWA Flight Center
The Beinecke Rare Book & Manuscript Library
Berliner Philharmonie
Leicester University Engineering Building
National Museum of Anthropology
Rudolph Hall
Salk Institute
High Court, Assembly, Chandigarh
Sea Ranch Condominium Complex
Piscinas de Marés
The Breuer Building (Whitney Museum)
Habitat '67
Xanadú, Calpe
Pilgrimage Church, Neviges
Art Center College of Design, Pasadena
Sheats Goldstein Residence
Nakagin Capsule Tower
Olympic Stadium, Munich
Phillips Exeter Academy Library
Kimbell Art Museum
Castelvecchio Museum
Casa Bianchi
Bagsværd Church
World Trade Center Towers
Centre Pompidou
Brion Family Tomb
Gehry House
Teatro del Mondo
Atheneum, New Harmony
Coxe-Hayden House and Studio
Vietnam Veterans Memorial
National Assembly Building, Bangladesh
Allied Bank Tower
Hongkong and Shanghai Bank Headquarters
National Museum of Roman Art
Church of the Light
Institut du Monde Arabe
Museum of Sculpture, São Paulo
Rooftop Remodeling Falkestrasse
Evergreen Building, Vancouver
Igualada Cemetery
Goetz Collection
Villa dall'Ava
Samitaur Tower
Therme Vals
Stretto House
6th Street House, Santa Monica
Marika-Alderton House
Santa Maria Church of Canaveses
Benesse House
Mesa Laboratory

沃尔夫·D·普瑞 / 蓝天组
(Wolf D. Prix / COOP HIMMELB(L)AU)

Monument to the Third International
German Pavilion (Barcelona Pavilion)
Philips Pavilion
Museu de Arte de São Paulo
Centre Le Corbusier (Heidi Weber House)
Centre Pompidou
Guggenheim Museum Bilbao
Austrian Cultural Forum
Jewish Museum Berlin
Vulcania Centre Européen du Volcanisme
Walt Disney Concert Hall
Seattle Central Library
Kanno Museum
BMW Welt
City of Culture of Galicia
Sifang Art Museum
CaixaForum Zaragoza
Heydar Aliyev Center
Musée des Confluences
Sagrada Família
Great Mosque of Djenné
Goetheanum
Chapelle Notre-Dame du Haut
Couvent Sainte-Marie de La Tourette
Cathedral of Brasília
Église Sainte-Bernadette du Banlay
Brion Family Tomb
Wotruba Church
Martin Luther Church, Hainburg
Olympic Stadium, Munich
Großes Schauspielhaus
Rusakov Workers' Club
Berliner Philharmonie
Sydney Opera House
Umbrella, Culver City
Dalian International Conference Center
Elbphilharmonie
Einstein Tower
Diamond Ranch High School
41 Cooper Square
Rolex Learning Center
Lovell Health House
Villa Savoye
Edgar J. Kaufmann House (Fallingwater)
Endless House
Stahl House (Case Study House No. 22)
Fuente de los Amantes
Das Canoas House
Chemosphere
Maison Drusch
Maison Bordeaux le Pecq
Studio Baumann, Vienna
Solohouse
Steinhaus, Lake Ossiach
Gehry House
Embryological House
L'Unité d'Habitation
Pedregulho Housing Complex
Edificio Copan
Habitat '67
Nakagin Capsule Tower
Torre Cube
Linked Hybrid
7132 Hotel & Arrival
Postsparkasse
Michaelerplatz House (Looshaus)
Loos' Tribune Tower Competition
Seagram Building
Salk Institute
BMW Hochhaus Vierzylinder
Bank of Georgia headquarters, Tbilisi
Zentralsparkasse Bank, Vienna
Hongkong and Shanghai Bank Headquarters
Pterodactyl, Culver City
CCTV Headquarters
Seat of the European Central Bank
Los Manantiales
Retti Candle Shop
Rooftop Remodeling Falkestrasse
Vitra Fire Station
Santa Caterina Market
Merida Factory Youth Movement
Kaiser Pavilion, Stadtbahn Station
National Congress, Brazil
Copacabana Breach Promenade
Radical Reconstruction in Havana
The City of the Future, Los Angeles 2106
TWA Flight Center
Druzhba Holiday Center Hall
The Golem (Film)
Wolkenbügel
New Babylon
The Sin Centre, London
Aircraft Carrier City in Landscape
Walking City
Plug-In City
Unterirdische Stadt
Wingnut, Stockholm
The Continuous Monument
Inhabiting the Quake

杰西·赖泽 + 梅本奈奈子
(Jesse Reiser + Nanako Umemoto)

Figini House
Villa Savoye
Chapelle Notre-Dame du Haut
Couvent Sainte-Marie de La Tourette
Villa Stein
The Assembly, Chandigarh
Villa Shodhan
Villa La Roche
Mill Owners' Association Building
L'Unité d'Habitation
Gallaratese II Apartments
Società Ippica Torinese
Casa Miller, Turin
TWA Flight Center
Cleveland Museum of Art, North Entrance
De Bijenkorf, Rotterdam
The Breuer Building (Whitney Museum)
Josephine Hagerty House
Casa Malaparte
Villa Mairea
Finlandia Hall
Paimio Sanatorium
National Congress, Brazil
Palácio do Planalto
Palácio da Alvorada
Frederick C. Robie House
Solomon R. Guggenheim Museum
S.C. Johnson & Son Headquarters
Edgar J. Kaufmann House (Fallingwater)
Unity Temple
Larkin Company Administration Building
Fagus Factory
Gropius House, Lincoln
Bauhaus Dessau
Weissenhof Estate, Haus 28-30
Hellerhof Estate
Einstein Tower
Hutfabrik Friedrich Steinberg, Herrmann & Co.
Schocken Department Store, Stuttgart
Villa Müller
Michaelerplatz House (Looshaus)
Villa Karma
Steiner House
American Bar, Vienna
Sulphuric Acid Factory, Lubon
Posen Tower
AEG Turbine Factory
Kreuzberg Tower
Berliner Philharmonie
Weissenhof Estate, Haus 33
German Pavilion (Barcelona Pavilion)
860–880 Lake Shore Drive
Seagram Building
Farnsworth House
Tugendhat House
Weissenhof Estate, Exhibition of 1927
Crown Hall
New National Gallery
Free University of Berlin
Salk Institute
Kimbell Art Museum
National Assembly Building, Bangladesh
Phillips Exeter Academy Library
Gwathmey Residence
House VI
Berlin Memorial to the Murdered Jews of Europe
City of Culture of Galicia
Smith House
Douglas House
Atheneum, New Harmony
High Museum of Art
Museum Angewandte Kunst, Frankfurt am Main
Jewish Museum Berlin
Centre Pompidou
The Menil Collection
The Shard
Shukhov Tower
Nakagin Capsule Tower
National Art Center, Tokyo
White U
Sendai Mediatheque
Stockholm Public Library
The Woodland Cemetery
Villa Snellman
Sky House, Tokyo
Edo Tokyo Museum
Miyakonojo Civic Hall
Rusakov Workers' Club
Melnikov House
La Maison de Verre
St. Mark's Church in Björkhagen
Church of St. Peter, Klippan
Palazzetto dello Sport
Zentralflughafen Tempelhof-Berlin
Goetheanum
Postsparkasse
Bank of London and South America
National Library of the Argentine Republic
Chiesa dell'Autostrada del Sole
Pilgrimage Church, Neviges

238 /

凯文·罗奇
（ Kevin Roche ）

Métropolitain Entrance
AEG Turbine Factory
Glasgow School of Art
Gamble House
Frederick C. Robie House
Casa Milà
Schindler House
Cité Frugès Housing Complex
Bauhaus Dessau
Lovell Health House
Stockholm Public Library
E-1027
German Pavilion (Barcelona Pavilion)
Chrysler Building
Villa Savoye
Paimio Sanatorium
Edgar J. Kaufmann House (Fallingwater)
S.C. Johnson & Son Headquarters
Villa Mairea
Dymaxion House
Baker House
Glass House
The Eames House (Case Study House No. 8)
Chapelle du Rosaire de Vence
Farnsworth House
L'Unité d'Habitation
Lever House
Muuratsalo Experimental House
Miller House, Columbus
Chapelle Notre-Dame du Haut
MIT Chapel
Sydney Opera House
Crown Hall
Maisons Jaoul
Seagram Building
Ingalls Rink
Solomon R. Guggenheim Museum
Church of the Three Crosses
Capilla de las Capuchinas
Couvent Sainte-Marie de La Tourette
US Air Force Academy Cadet Chapel
TWA Flight Center
The Beinecke Rare Book & Manuscript Library
Gateway Arch
John Deere World Headquarters
Marina City, Chicago
Olivetti Factory, Merlo
Rudolph Hall
Salk Institute
The Assembly, Chandigarh
Kuwait Embassy, Tokyo
The Breuer Building (Whitney Museum)
Habitat '67
Smith House
Hyatt Regency, Atlanta
Orange County Government Centre, Goshen
Ford Foundation Headquarters
Knights of Columbus Building
Montreal Biosphère
National Assembly Building, Bangladesh
Charles De Gaulle Airport, Terminal 1 & 2A–2F
House III (Miller House)
Aula Paolo VI
Phillips Exeter Academy Library
Kimbell Art Museum
Robin Hood Gardens
Sears Tower
Marquette Plaza
Casa Bianchi
House VI
Hirshhorn Museum
Whig Hall, Princeton University
Prentice Women's Hospital Building
1 United Nations Plaza
Bagsværd Church
World Trade Center Towers
Centre Pompidou
Gehry House
Crystal Cathedral
Koshino House
Vietnam Veterans Memorial
Vitra Design Museum
Zurichtoren
Lloyd's of London
The Menil Collection
Pyramide du Louvre
Reykjavík Town Hall
Glass Chapel, Alabama
Vitra Fire Station
Paper Church, Kobe
Guggenheim Museum Bilbao
Getty Center
Diamond Ranch High School
Church of the Light
London Millennium Footbridge
Tate Modern
Nine Bridges Country Club
Jewish Museum Berlin
American Folk Art Museum
Petronas Towers

理查德·罗杰斯爵士
（ Sir Richard Rogers ）

AEG Turbine Factory
Glasgow School of Art
Gamble House
Frederick C. Robie House
Posen Tower
Postsparkasse
Schindler House
Schröderhuis
Cité Frugès Housing Complex
Lovell Beach House
Bauhaus Dessau
Lovell Health House
Stockholm Public Library
German Pavilion (Barcelona Pavilion)
Rusakov Workers' Club
Villa Savoye
La Maison de Verre
Haus Schminke
Edgar J. Kaufmann House (Fallingwater)
S.C. Johnson & Son Headquarters
Villa Mairea
The Woodland Cemetery
Dymaxion House
Baker House
Barragán House and Studio
Turin Exhibition Hall
Glass House
The Eames House (Case Study House No. 8)
Farnsworth House
L'Unité d'Habitation
Lever House
Pedregulho Housing Complex
Muuratsalo Experimental House
Cité d'Habitacion of Carrières Centrales
Chapelle Notre-Dame du Haut
Sydney Opera House
Crown Hall
Maisons Jaoul
Mill Owners' Association Building
Torre Velasca
Seagram Building
Museum of Modern Art, Rio de Janeiro
National Congress, Brazil
Solomon R. Guggenheim Museum
Church of the Three Crosses
House Van den Doel
Taliesin West
Stahl House (Case Study House No. 22)
Couvent Sainte-Marie de La Tourette
Halen Estate
TWA Flight Center
Snowdon Aviary
Berliner Philharmonie
Leicester University Engineering Building
National Olympic Gymnasium, Tokyo
Otaniemi Technical University
Peabody Terrace
Salk Institute
The Assembly, Chandigarh
Smith House
Ford Foundation Headquarters
Banco Ciudad de Buenos Aires (Casa Matriz)
Art Center College of Design, Pasadena
House III (Miller House)
Nakagin Capsule Tower
Olympic Stadium, Munich
Phillips Exeter Academy Library
Kimbell Art Museum
Castelvecchio Museum
Marquette Plaza
Sainsbury Center for Visual Arts
Centre Pompidou
Atheneum, New Harmony
Koshino House
Vietnam Veterans Memorial
National Assembly Building, Bangladesh
Lloyd's of London
Hongkong and Shanghai Bank Headquarters
Guggenheim Museum Bilbao
Diamond Ranch High School
Millennium Dome
Bordeaux Law Courts
Sendai Mediatheque
Highpoint Apartment Blocks
John Hancock Center
Casa Milà
Stadio San Nicola
E-1027
Maison Jean Prouvé, Nancy
Centre Le Corbusier (Heidi Weber House)
Adolfo Madrid–Barajas Airport
The Leadenhall Building

摩西·萨夫迪
(Moshe Safdie)

AEG Turbine Factory
Glasgow School of Art
Grand Palais
Gamble House
Frederick C. Robie House
Postsparkasse
Schindler House
Fiat Works
Schröderhuis
Bauhaus Dessau
Lovell Health House
Stockholm Public Library
Villa Müller
Einstein Tower
German Pavilion (Barcelona Pavilion)
Villa Savoye
La Maison de Verre
Edgar J. Kaufmann House (Fallingwater)
S.C. Johnson & Son Headquarters
Hadassah Medical Center, Mount Scopus
New Gourna Village Mosque
Barragán House and Studio
Dymaxion House
Turin Exhibition Hall
The Eames House (Case Study House No. 8)
Farnsworth House
L'Unité d'Habitation
Lever House
Cité d'Habitacion of Carrières Centrales
Highlife Textile Factory
Chapelle Notre-Dame du Haut
Sydney Opera House
Crown Hall
Maisons Jaoul
Mill Owners' Association Building
Seagram Building
National Congress, Brazil
Solomon R. Guggenheim Museum
Church of the Three Crosses
Sabarmati Ashram
Stahl House (Case Study House No. 22)
Fredensborg Housing
Couvent Sainte-Marie de La Tourette
Municipal Orphanage
TWA Flight Center
Snowdon Aviary
Querini Stampalia Renovation
The Beinecke Rare Book & Manuscript Library
John Deere World Headquarters
Berliner Philharmonie
National Olympic Gymnasium, Tokyo
Shrine of the Book
Salk Institute
The Assembly, Chandigarh
Sea Ranch Condominium Complex
Habitat '67
Smith House
Hyatt Regency, Atlanta
Ford Foundation Headquarters
Banco Ciudad de Buenos Aires (Casa Matriz)
New National Gallery
Flower Kiosk, Malmö Cemetery
Montreal Biosphère
Israel Museum
Piscinas de Marés
Golden Mile Complex
Nakagin Capsule Tower
Olympic Stadium, Munich
Phillips Exeter Academy Library
Kimbell Art Museum
Castelvecchio Museum
Casa Bianchi
House VI
Centraal Beheer Building
Centre Pompidou
Brion Family Tomb
Gehry House
Museum of Anthropology, UBC
National Gallery of Art, East Building
National Gallery of Canada
Vietnam Veterans Memorial
National Assembly Building, Bangladesh
Neue Staatsgalerie
Lloyd's of London
Hongkong and Shanghai Bank Headquarters
National Museum of Roman Art
Pyramide du Louvre
Sri Lankan Parliament Building
Water Temple, Hyogo
Beyeler Foundation
The Menil Collection
Skirball Cultural Center
Yad Vashem
Guggenheim Museum Bilbao
Bordeaux Law Courts
Sendai Mediatheque
DZ Bank Building
Therme Vals
Portuguese National Pavillion
British Museum

斯坦利·塞陶维兹
(Stanley Saitowitz)

Larkin Company Administration Building
Unity Temple
Schindler House
Fiat Works
Ozenfant House
Villa Cook
Van Nelle Factory
Schröderhuis
Cité Frugès Housing Complex
Soviet Pavilion
Shukhov Tower
Lovell Beach House
Bauhaus Dessau
Villa Stein
Lovell Health House
Weissenhof Estate, Exhibition of 1927
Melnikov House
German Pavilion (Barcelona Pavilion)
Rusakov Workers' Club
Narkomfin Building
Lange and Esters Houses
Tugendhat House
Villa Savoye
Paimio Sanatorium
La Maison de Verre
Molitor Building
Edgar J. Kaufmann House (Fallingwater)
Casa del Fascio
S.C. Johnson & Son Headquarters
Villa Mairea
The Woodland Cemetery
Aarhus City Hall
Baker House
Barragán House and Studio
Glass House
The Eames House (Case Study House No. 8)
Architect's Second House, São Paulo
Farnsworth House
Das Canoas House
Säynätsalo Town Hall
L'Unité d'Habitation
Lever House
Pedregulho Housing Complex
Hunstanton School
Chapelle Notre-Dame du Haut
The Assembly, Chandigarh
St. Mark's Church in Björkhagen
Villa Sarabhai
Sydney Opera House
Crown Hall
Maisons Jaoul
Mill Owners' Association Building
Studio Aalto
House of the Future, 1956
Capilla de las Capuchinas
Richards Medical Research Laboratories
Seagram Building
National Congress, Brazil
Unité d'Habitation, Berlin
Pepsi-Cola Building
Stahl House (Case Study House No. 22)
Couvent Sainte-Marie de La Tourette
Municipal Orphanage
Esherick House
Gimansio del Colegio Maravillas
Carpenter Center for the Visual Arts
Leicester University Engineering Building
Casa Butantã
Frey House II
Rudolph Hall
Vanna Venturi House
The Economist Building, London
Centre Le Corbusier (Heidi Weber House)
Salk Institute
Transvaal House
Piscinas de Marés
Habitat '67
Norman Fisher House
New National Gallery
Kappe Residence
FAU, University of São Paulo
Toronto-Dominion Centre
Art Center College of Design, Pasadena
National Assembly Building, Bangladesh
Indian Institute of Management, Ahmedabad
San Cataldo Cemetery
Phillips Exeter Academy Library
Franklin D. Roosevelt Four Freedoms Park
Kimbell Art Museum
Robin Hood Gardens
Free University of Berlin
Gallaratese II Apartments
Yale Center for British Art
Bagsvaerd Church
Centre Pompidou
Viipuri Municipal Library
Institut du Monde Arabe
Museum of Sculpture, São Paulo
Barcelona Olympic Archery Range
Dominus Winery

麦克·斯科金
(Mack Scogin)

Palais Stoclet
Bavinger House
Stockholm Public Library
Skandia Theater
The Woodland Cemetery
Koshino House
Naoshima Contemporary Art Museum
Bauhaus Dessau
Baker House
Paimio Sanatorium
Iberê Camargo Foundation
Boa Nova Tea House
Piscinas de Marés
Faculty of Architecture of the University of Porto
Schröderhuis
Einstein Tower
S.C. Johnson & Son Headquarters
Edgar J. Kaufmann House (Fallingwater)
Taliesin West
Solomon R. Guggenheim Museum
Haus Schminke
Berliner Philharmonie
Berlin State Library
Gehry House
Guggenheim Museum Bilbao
Villa Savoye
Chapelle Notre-Dame du Haut
Mill Owners' Association Building
Villa Shodhan
Rockefeller Center
La Maison de Verre
Tristan Tzara House
American Bar, Vienna
Villa Müller
Goetheanum
Casa Milà
Flower Kiosk, Malmö Cemetery
St. Mark's Church in Björkhagen
Malmö Eastern Cemetery
National Gallery, Sainsbury Wing
Vanna Venturi House
Centre Pompidou
Neue Staatsgalerie
Wexner Center for the Visual Arts
Hongkong and Shanghai Bank Headquarters
Lloyd's of London
Portland Building
Sheats Goldstein Residence
Chemosphere
Elrod House
Arango-Marbrisa House
Levy Residence, Malibu
Villa dall'Ava
CCTV Headquarters
Casa da Música
Seattle Central Library
Brion Family Tomb
Museo Canova
Castelvecchio Museum
Igualada Cemetery
Scottish Parliament Building
Casa del Fascio
Asilo Sant'Elia
Rusakov Workers' Club
Das Canoas House
The Eames House (Case Study House No. 8)
Rudolph Residence, 23 Beekman Place
Rudolph Hall
London Zoo Penguin Pool
Church of the Most Sacred Heart of Our Lord
CBS Building, New York
TWA Flight Center
Gateway Arch
Ingalls Rink
Schindler House
Salk Institute
Kimbell Art Museum
Esherick House
National Assembly Building, Bangladesh
German Pavilion (Barcelona Pavilion)
Seagram Building
Farnsworth House
Toronto-Dominion Centre
Kluczynski Federal Building
Vitra Fire Station
John Hancock Tower
Beijing National Stadium
Tate Modern
CaixaForum Madrid
Atlanta Marriott Marquis
Hyatt Regency, Atlanta
The Breuer Building (Whitney Museum)
St. John's Abbey, Collegeville
UC Campus Recreation Center
Emerson College Los Angeles
Selfridges Birmingham
Habitat '67
Barragán House and Studio
Glass Chapel, Alabama
Guthrie Theater

妹岛和世 + 西泽立卫 /SANNA 事务所
(Kazuyo Sejima + Ryue Nishizawa / SANAA)

Larkin Company Administration Building
Glasgow School of Art
Gamble House
Frederick C. Robie House
Casa Milà
Michaelerplatz House (Looshaus)
Postsparkasse
Park Güell
Schindler House
Fiat Works
Schröderhuis
Bauhaus Dessau
Lovell Beach House
Stockholm Public Library
Lovell Health House
German Pavilion (Barcelona Pavilion)
Melnikov House
Tugendhat House
Villa Müller
Villa Savoye
Van Nelle Factory
La Maison de Verre
Edgar J. Kaufmann House (Fallingwater)
Casa del Fascio
S.C. Johnson & Son Headquarters
Villa Mairea
The Woodland Cemetery
Casa Malaparte
Dymaxion House
Barragán House and Studio
Baker House
The Eames House (Case Study House No. 8)
Glass House
Farnsworth House
Säynätsalo Town Hall
L'Unité d'Habitation
Yale University Art Gallery
Ibirapuera Park
Chapelle Notre-Dame du Haut
Villa Sarabhai
Trenton Bath House
Crown Hall
Mill Owners' Association Building
860–880 Lake Shore Drive
Seagram Building
Sky House, Tokyo
Solomon R. Guggenheim Museum
National Congress, Brazil
Couvent Sainte-Marie de La Tourette
Municipal Orphanage, Amsterdam
Stahl House (Case Study House No. 22)
TWA Flight Center
Carpenter Center for the Visual Arts
Berliner Philharmonie
Leicester University Engineering Building
The Beinecke Rare Book & Manuscript Library
Querini Stampalia Renovation
Rudolph Hall
National Olympic Gymnasium, Tokyo
Kagawa Prefectural Government Hall
Salk Institute
The Assembly, Chandigarh
Centre Le Corbusier (Heidi Weber House)
Piscinas de Marés
Habitat '67
Smith House
Montreal Biosphère
Hillside Terrace Complex I–VI
Ford Foundation Headquarters
New National Gallery
Museu de Arte de São Paulo
Toronto-Dominion Centre
Nakagin Capsule Tower
Kimbell Art Museum
Phillips Exeter Academy Library
Olympic Stadium, Munich
Castelvecchio Museum
Free University of Berlin
Centraal Beheer Building
Yale Center for British Art
Bagsværd Church
Centre Pompidou
SESC Pompéia
Brion Family Tomb
National Assembly Building, Bangladesh
Neue Staatsgalerie
Hongkong and Shanghai Bank Headquarters
Lloyd's of London
National Museum of Roman Art
Igualada Cemetery
Parc de la Villette
Santa Maria Church de Canaveses
Maison à Bordeaux
Vitra Fire Station
Guggenheim Museum Bilbao
Diamond Ranch High School
Jewish Museum Berlin
Sendai Mediatheque
Yokohama International Port Terminal
Ray and Maria Stata Center

豪尔郝·西尔韦蒂
（Jorge Silvetti）

Beurs van Berlage
Larkin Company Administration Building
Frederick C. Robie House
Casa Milà
Palais Stoclet
Park Güell
Ozenfant House
Schröderhuis
Villa Stein
Karl-Marx-Hof
Stockholm Public Library
Villa Moller
German Pavilion (Barcelona Pavilion)
Villa Müller
Tugendhat House
Villa Savoye
La Maison de Verre
Frontón Recoletos
Edgar J. Kaufmann House (Fallingwater)
S.C. Johnson & Son Headquarters
Villa Mairea
Arve Bridge
The Woodland Cemetery
Casa Malaparte
Casa del Puente
Baker House
Barragán House and Studio
The Eames House (Case Study House No. 8)
Farnsworth House
Säynätsalo Town Hall
L'Unité d'Habitation
Lever House
Pedregulho Housing Complex
Muuratsalo Experimental House
Chapelle Notre-Dame du Haut
The Assembly, Chandigarh
Sydney Opera House
Maisons Jaoul
St. Mark's Church in Björkhagen
Palazzetto dello Sport
Torre Velasca
Seagram Building
National Congress, Brazil
Solomon R. Guggenheim Museum
Church of Cristo Obrero
Pepsi-Cola Building
Couvent Sainte-Marie de La Tourette
Toulouse-le-Mirail Housing
US Air Force Academy Cadet Chapel
TWA Flight Center
The Beinecke Rare Book & Manuscript Library
John Deere World Headquarters
Berliner Philharmonie
Leicester University Engineering Building
National Olympic Gymnasium, Tokyo
National Museum of Anthropology
Peabody Terrace
Vanna Venturi House
The Economist Building, London
Salk Institute
Sea Ranch Condominium Complex
Piscinas de Marés
Kuwait Embassy, Tokyo
Edificio Copan
The Breuer Building (Whitney Museum)
Bank of London and South America
Habitat '67
Hyatt Regency, Atlanta
Torre Blancas
Ford Foundation Headquarters
History Faculty Building, Cambridge
Olivetti Training Center
Itamaraty Palace
Nakagin Capsule Tower
Olympic Stadium, Munich
Phillips Exeter Academy Library
Snyderman House
Kimbell Art Museum
Castelvecchio Museum
Free University of Berlin
John Hancock Tower
Gallaratese II Apartments
Centre Pompidou
House VI
Gehry House
Teatro del Mondo
Atlantis Condominium
Neue Staatsgalerie
National Museum of Roman Art
The Menil Collection
Igualada Cemetery
Goetz Collection
Maison à Bordeaux
Dominus Winery
Therme Vals
Educatorium
Guggenheim Museum Bilbao
Sendai Mediatheque
Yokohama International Port Terminal
DZ Bank Building

安德烈娅·西米奇 + 瓦尔·K·沃克
（Andrea Simitch + Val K. Warke）

Säynätsalo Town Hall
Villa Mairea
Gund Hall
Atlantis Condominium
Architect's Second House, São Paulo
Stockholm Public Library
The Woodland Cemetery
Cuadra San Cristóbal
AEG Turbine Factory
Museu de Arte de São Paulo
Casa Bianchi
The Breuer Building (Whitney Museum)
Bacardi Bottling Plant
La Maison de Verre
Rooftop Remodeling Falkestrasse
Karl-Marx-Hof
House II (Vermont House)
House VI
New Gourna Village Mosque
Center for Hydrographic Studies, Madrid
Hongkong and Shanghai Bank Headquarters
Montreal Biosphère
Il Bagno de Bellinzona
Gehry House
Guggenheim Museum Bilbao
Hanselmann House
E-1027
Gamble House
Bauhaus Dessau
Vitra Fire Station
Ricola Storage Building
Storefront for Art and Architecture
Salk Institute
Kimbell Art Museum
Stahl House (Case Study House No. 22)
Nakagin Capsule Tower
Chapelle Notre-Dame du Haut
Villa Savoye
Villa Stein
Couvent Sainte-Marie de La Tourette
L'Unité d'Habitation
Carpenter Center for the Visual Arts
The Assembly, Chandigarh
Church of St. Peter, Klippan
Casa Malaparte
Villa Müller
Glasgow School of Art
Smith House
Atheneum, New Harmony
Soviet Pavilion
Schocken Department Store, Stuttgart
Museum of Sculpture, São Paulo
Seagram Building
Farnsworth House
German Pavilion (Barcelona Pavilion)
New National Gallery
Crown Hall
National Museum of Roman Art
Igualada Cemetery
Casa Il Girasole
Diamond Ranch High School
Villa VPRO
Palazzetto dello Sport
Lovell Health House
National Congress, Brazil
Institut du Monde Arabe
Kunsthal, Rotterdam
Educatorium
Stockholm City Hall
Rue Franklin Apartments
Centre Pompidou
National and University Library of Slovenia
Villa Planchart
Teatro del Mondo
Gallaratese II Apartments
Rudolph Hall
MIT Chapel
Habitat '67
Berliner Philharmonie
Lovell Beach House
Peabody Terrace
Piscinas de Marés
Galician Center of Contemporary Art
Hunstanton School
Lever House
The Beinecke Rare Book & Manuscript Library
House at the Bom Jesus, Braga
Leicester University Engineering Building
History Faculty Building, Cambridge
Neue Staatsgalerie
Reykjavík Town Hall
Casa del Fascio
Asilo Sant'Elia
Fiat Works
Parc de la Villette
T-House, Wilton
Can Lis, Majorca
Postsparkasse
Edgar J. Kaufmann House (Fallingwater)
Solomon R. Guggenheim Museum

罗伯特·A·M·斯特恩
(Robert A.M. Stern)

Chapelle Notre-Dame du Haut
Couvent Sainte-Marie de La Tourette
Villa Savoye
Farnsworth House
Glass House
TWA Flight Center
Guggenheim Museum Bilbao
S.C. Johnson & Son Headquarters
Leicester University Engineering Building
La Maison de Verre
Edgar J. Kaufmann House (Fallingwater)
Seagram Building
Postsparkasse
Salk Institute
The Woodland Cemetery
Kimbell Art Museum
AEG Turbine Factory
Lovell Beach House
Solomon R. Guggenheim Museum
Rudolph Hall
Stockholm Public Library
Frederick C. Robie House
Phillips Exeter Academy Library
Villa Mairea
Ford Foundation Headquarters
Baker House
Crown Hall
Glasgow School of Art
Vanna Venturi House
Neue Staatsgalerie
National Museum of Roman Art
Larkin Company Administration Building
Carpenter Center for the Visual Arts
Casa Milà
Park Güell
Karl-Marx-Hof
Tugendhat House
Villa Müller
Van Nelle Factory
Casa Il Girasole
Säynätsalo Town Hall
Sea Ranch Condominium Complex
Posen Tower
Beurs van Berlage
New National Gallery
Miller House, Columbus
Gamble House
Hilversum Town Hall
Rockefeller Center
Torre Velasca
Gordon Wu Hall
Unity Temple
Schocken Department Store, Stuttgart
MIT Chapel
John Deere World Headquarters
Knights of Columbus Building
National Farmers Bank
Skandia Theater
Villa La Roche
Castle Drogo
Yale Center for British Art
Imperial Hotel, Tokyo
Chrysler Building
National Museum of Anthropology
Kresge College
United Nations Headquarters
American Bar, Vienna
Großes Schauspielhaus
Stockholm City Hall
Worker's Housing Estate, Hoek van Holland
Cranbrook Academy of Art
Memorial Quadrangle, Yale University
Barclay-Vesey Building
General Motors Technical Center
Hirsch Residence
Taliesin West
Taliesin East
Town Hall, Logroño
National Gallery of Art, Washington, D.C.
Kirche am Steinhof
Villa Kerylos
PSFS Building
National and University Library of Slovenia
Ingalls Rink
Dixwell Fire Station
Bass Residence
National Gallery, Sainsbury Wing
Rashtrapati Bhavan
Union Station, Washington, D.C.
Helsinki Central Railway Station
Stuttgart Hauptbahnhof
Grand Central Terminal
Lincoln Memorial
Chicago Tribune Tower
Grundtvig Memorial Church
Église Notre-Dame du Raincy
Garbatella Quarter
Pasadena City Hall
Arco della Vittoria
Triple Bridge

安娜·托斯通伊斯
(Ana Tostões)

Studio Aalto
AEG Turbine Factory
Rádio Nacional de Angola
Postsparkasse
Barragán House and Studio
Bauhaus Dessau
Beira Railway Station
Berliner Philharmonie
Boa Nova Tea House
Estádio Municipal de Braga
Museum of Sculpture, São Paulo
Brion Family Tomb
Casa Butantã
Calouste Gulbenkian Foundation
Das Canoas House
Casa de Vidro
Casa del Fascio
Castelvecchio Museum
Edificio de CEPAL
The Assembly, Chandigarh
Chapelle Notre-Dame du Haut
Hotel Chuabo
Le Lignon
Gimansio del Colegio Maravillas
Couvent Sainte-Marie de La Tourette
Dymaxion House
Edgar J. Kaufmann House (Fallingwater)
FAU, University of São Paulo
Farnsworth House
Fiat Works
Bank of London and South America
Guggenheim Museum Bilbao
Hizuchi Elementary School
Hotel do Mar, Sesimbra
Institut du Monde Arabe
International House of Japan
Pueblo de colonización El Realengo
Kagawa Prefectural Government Hall
Kimbell Art Museum
L'Unité d'Habitation
Los Manantiales
Lovell Beach House
Casa Malaparte
Michaelerplatz House (Looshaus)
Gustavo Capanema Palace
Olympic Stadium, Munich
Municipal Orphanage, Amsterdam
Niterói Contemporary Art Museum
Museum of Modern Art, Rio de Janeiro
Nakagin Capsule Tower
National Assembly Building, Bangladesh
National Congress, Brazil
National Museum of Roman Art
National Olympic Gymnasium, Tokyo
New Gourna Village Mosque
New National Gallery
Nissay Theatre
Cathedral of Brasília
Quinta da Conceição
Residência Tomie Ohtake
Rusakov Workers' Club
Crown Hall
Salk Institute
Museu de Arte de São Paulo
Säynätsalo Town Hall
Schindler House
Schröderhuis
Seagram Building
Sendai Mediatheque
SESC Pompéia
Shiba Ryotaro Memorial Museum
Halen Estate
Hufeisensiedlung
The Woodland Cemetery
Smiling Lion Apartment Building
Stahl House (Case Study House No. 22)
Stockholm Public Library
Sydney Opera House
Civil Government Building, Tarragona
The Breuer Building (Whitney Museum)
The Eames House (Case Study House No. 8)
Einstein Tower
Hillside Terrace Complex I–VI
St. Mark's Church in Björkhagen
Museum of Modern Art, Kamakura & Hayama
National Museum of Western Art
Resurrection Chapel, Turku
Therme Vals
Piscinas de Marés
Tugendhat House
Turin Exhibition Hall
Casa Ugalde
Van Nelle Factory
Viipuri Municipal Library
Architect's Second House, São Paulo
Villa Mairea
Villa Müller
Villa Savoye
Weissenhof Estate, Exhibition of 1927
Yokohama International Port Terminal

伯纳德·屈米
(Bernard Tschumi)

Beurs van Berlage
AEG Turbine Factory
Postsparkasse
Glasgow School of Art
Rue Franklin Apartments
Casa Milà
Fagus Factory
Glass Pavilion
Michaelerplatz House (Looshaus)
Großes Schauspielhaus
Fiat Works
Grand Central Terminal
Narkomfin Building
Open-air School, Amsterdam
Rusakov Workers' Club
Einstein Tower
Schindler House
Schröderhuis
Kiefhoek Housing Development
Co-op Zimmer
Hufeisensiedlung
Stockholm Public Library
Paimio Sanatorium
Karl-Marx-Hof
Villa Savoye
Casa del Fascio
Rockefeller Center
La Maison de Verre
S.C. Johnson & Son Headquarters
Casa Malaparte
London Zoo Penguin Pool
The Eames House (Case Study House No.8)
L'Unité d'Habitation
Farnsworth House
Glass House
Casa del Puente
Turin Exhibition Hall
Municipal Orphanage
Nestlé Headquarters
Robin Hood Gardens
Sydney Opera House
Seagram Building
Salk Institute
Leicester University Engineering Building
Los Manantiales
Chapelle Notre-Dame du Haut
National Congress, Brazil
Solomon R. Guggenheim Museum
Snowdon Aviary
Free University of Berlin
Berliner Philharmonie
Centraal Beheer Building
Église Sainte-Bernadette du Banlay
TWA Flight Center
Carpenter Center for the Visual Arts
Oita Prefectural Library
Gallaratese II Apartments
German Pavilion, Expo '67
Montreal Biosphère
Rudolph Hall
Vanna Venturi House
Smith House
Shrine of the Book
Habitat '67
Ford Foundation Headquarters
Museu de Arte de São Paulo
Umlauftank
Nakagin Capsule Tower
San Cataldo Cemetery
Hirshhorn Museum
Westin Bonaventure Hotel
Wall House II
Willis Faber and Dumas Headquarters
Centre Pompidou
Gehry House
Beires House
White U
House III (Miller House)
World Trade Center Towers
Byker Wall
La Fábrica, Sant Just Desvern
Neue Staatsgalerie
Church of the Light
Netherlands Dance Theatre, Hague
Parc de la Villette
Wexner Center for the Visual Arts
Institut du Monde Arabe
National Museum of Roman Art
Rooftop Remodeling Falkestrasse
Goetz Collection
Diamond Ranch High School
Expo 2000, Netherlands Pavilion
Kunsthal, Rotterdam
Kiasma
Guggenheim Museum Bilbao
Vitra Fire Station
Le Fresnoy Art Center
Igualada Cemetery
Jewish Museum Berlin
Sendai Mediatheque

本·范贝克尔 + 卡罗琳·博斯 /UNS 工作室
(Ben van Berkel + Caroline Bos / UNStudio)

Maison & Atelier Horta
Beurs van Berlage
Larkin Company Administration Building
Riehl House
AEG Turbine Factory
Casa Milà
Posen Tower
Michaelerplatz House (Looshaus)
Park Güell
Glass Pavilion
Großes Schauspielhaus
Einstein Tower
Soviet Pavilion
Villa Stein
Goetheanum
Rusakov Workers' Club
Villa Savoye
La Maison de Verre
Haus Schminke
Villa Girasole
Orvieto Aircraft Hangars
Casa del Fascio
S.C. Johnson & Son Headquarters
Casa Malaparte
Palazzo della Civiltà Italiana
Baker House
Turin Exhibition Hall
Säynätsalo Town Hall
Casa de Vidro
Gatti Wool Factory
Iglesia de la Medalla Milagrosa
Catalano House
Chapelle Notre-Dame du Haut
Hiroshima Peace Memorial Museum
Corso Italia Complex, Milan
Ginásio do Clube Atlético Paulistano
Palazzetto dello Sport
National Congress, Brazil
Los Manantiales
Palazzo del Lavoro
Couvent Sainte-Marie de La Tourette
Municipal Orphanage
National Library of the Argentine Republic
TWA Flight Center
Leicester University Engineering Building
Berliner Philharmonie
National Olympic Gymnasium, Tokyo
Olivetti Factory, Merlo
Rudolph Hall
Kuwait Embassy, Tokyo
Oita Prefectural Library
Sonsbeek Pavilion
Bank of London and South America
Musmeci Bridge
Shizuoka Press and Broadcasting Center
Habitat '67
Smith House
Kowloon Walled City
Pilgrimage Church, Neviges
Museu de Arte de São Paulo
Mivtachim Sanitarium
FAU, University of São Paulo
Itamaraty Palace
House II (Vermont House)
House III (Miller House)
White U
Nakagin Capsule Tower
Kimbell Art Museum
House VI
Museum of Modern Art, Gunma
Teatro Regio
Town Hall, Logroño
Sumiyoshi Row House
SESC Pompéia
Centre Pompidou
Atheneum, New Harmony
Koshino House
Venice III House
Lloyd's of London
Hongkong and Shanghai Bank Headquarters
National Museum of Roman Art
Ricola Storage Building
The Menil Collection
Church of the Light
Institut du Monde Arabe
TEPIA Science Pavilion, Tokyo
Vitra Design Museum
Erasmus Bridge
Parc de la Villette
T-House, Wilton
House at the Bom Jesus, Braga
Villa Wilbrink
Mobius House
Galician Center of Contemporary Art
Vitra Fire Station
Therme Vals
Villa VPRO
Bordeaux Law Courts
Yokohama International Port Terminal
Miyagi Stadium

罗伯特·文丘里 + 丹尼斯·斯科特·布朗
(Robert Venturi + Denise Scott Brown)

Postsparkasse
Schindler House
Fiat Works
Schröderhuis
Cité Frugès Housing Complex
Karl-Marx-Hof
Lovell Health House
Stockholm Public Library
Villa Moller
German Pavilion (Barcelona Pavilion)
Villa Savoye
Van Nelle Factory
La Maison de Verre
Villa Girasole
Edgar J. Kaufmann House (Fallingwater)
Casa del Fascio
The Woodland Cemetery
Casa Malaparte
Dymaxion House
Turin Exhibition Hall
Glass House
The Eames House (Case Study House No. 8)
Baker House
Muuratsalo Experimental House
Farnsworth House
L'Unité d'Habitation
Lever House
Pedregulho Housing Complex
Crown Hall
Maisons Jaoul
Chapelle Notre-Dame du Haut
Seagram Building
National Congress, Brazil
Philips Pavilion
Solomon R. Guggenheim Museum
Cité d'Habitacion of Carrières Centrales
Stahl House (Case Study House No. 22)
Couvent Sainte-Marie de La Tourette
Municipal Orphanage
Halen Estate
TWA Flight Center
Querini Stampalia Renovation
Berliner Philharmonie
Leicester University Engineering Building
Vanna Venturi House
National Olympic Gymnasium, Tokyo
Peabody Terrace
Rudolph Hall
Salk Institute
The Assembly, Chandigarh
Centre Le Corbusier (Heidi Weber House)
Piscinas de Marés
Kuwait Embassy, Tokyo
Bank of London and South America
Habitat '67
Smith House
Museu de Arte de São Paulo
Ford Foundation Headquarters
Kowloon Walled City
Benacerraf House Project
Flower Kiosk, Malmö Cemetery
Petrobras Headquarters
The Beinecke Rare Book & Manuscript Library
House III (Miller House)
Nakagin Capsule Tower
Olympic Stadium, Munich
Phillips Exeter Academy Library
Kimbell Art Museum
Free University of Berlin
Byker Wall
Centraal Beheer Building
Whig Hall, Princeton University
Bagsværd Church
Brion Family Tomb
Centre Pompidou
SESC Pompéia
Hedmark Cathedral Museum
Gehry House
Atheneum, New Harmony
Koshino House
Vietnam Veterans Memorial
National Assembly Building, Bangladesh
Neue Staatsgalerie
Lloyd's of London
Hongkong and Shanghai Bank Headquarters
Rooftop Remodeling Falkestrasse
National Museum of Roman Art
Void Space/Hinged Space Housing
Igualada Cemetery
Parc de la Villette
Maison à Bordeaux
Samitaur Tower
Guggenheim Museum Bilbao
UFA Cinema Centre
Jewish Museum Berlin
Bordeaux Law Courts
Diamond Ranch High School
Yokohama International Port Terminal
Sendai Mediatheque
Empty Sky

马里昂·韦斯 + 迈克尔·曼弗雷迪
(Marion Weiss + Michael Manfredi)

Greywalls
Postsparkasse
Glasgow School of Art
Frederick C. Robie House
Park Güell
Schindler House
Lovell Beach House
Villa Stein
Stockholm Public Library
Lovell Health House
Tugendhat House
Villa Savoye
German Pavilion (Barcelona Pavilion)
Triple Bridge
Van Nelle Factory
Rockefeller Center
La Maison de Verre
City of Refuge, Paris
Casa Rustici
S.C. Johnson & Son Headquarters
Casa del Fascio
Taliesin West
Edgar J. Kaufmann House (Fallingwater)
Strathmore Apartments
Casa Malaparte
Villa Mairea
The Woodland Cemetery
L'Unité d'Habitation
Kaufmann House
Farnsworth House
Barragán House and Studio
Baker House
Casa Il Girasole
Crown Hall
Säynätsalo Town Hall
Maisons Jaoul
The Assembly, Chandigarh
Capilla de las Capuchinas
Chapelle Notre-Dame du Haut
Sydney Opera House
Mill Owners' Association Building
Olivetti Showroom
Church of the Three Crosses
Couvent Sainte-Marie de La Tourette
House of Culture
Seagram Building
United States Embassy, New Delhi
Rudolph Hall
Solomon R. Guggenheim Museum
Leicester University Engineering Building
Castelvecchio Museum
Church of Cristo Obrero
Salk Institute
TWA Flight Center
Carpenter Center for the Visual Arts
Querini Stampalia Renovation
Centre Le Corbusier (Heidi Weber House)
George Washington Bridge Bus Station
Free University of Berlin
National Olympic Gymnasium, Tokyo
Marina City, Chicago
Andrew Melville Hall
Piscinas de Marés
Norman Fisher House
Smith House
Habitat '67
Museu de Arte de São Paulo
Ford Foundation Headquarters
New National Gallery
Gallaratese II Apartments
FAU, University of São Paulo
Hillside Terrace Complex I–VI
House III (Miller House)
Penn Mutual Tower
Centre Pompidou
Saint-Pierre, Firminy
Phillips Exeter Academy Library
Kimbell Art Museum
Centraal Beheer Building
Yale Center for British Art
Casa Gilardi
Bagsværd Church
Neue Staatsgalerie
Gehry House
University of Urbino
National Museum of Roman Art
Parliament House, Canberra
National Assembly Building, Bangladesh
Parc de la Villette
The Menil Collection
Saint Benedict Chapel, Sumvitg
Igualada Cemetery
Kunsthal, Rotterdam
Vitra Fire Station
Chapel of St. Ignatius
Guggenheim Museum Bilbao
Sendai Mediatheque
Diamond Ranch High School
Nakagin Capsule Tower
Serralves Museum of Contemporary Art

托德·威廉姆斯 + 比利·钱
(Tod Williams + Billie Tsien)

Beurs van Berlage
Rue Franklin Apartments
AEG Turbine Factory
Glasgow School of Art
Gamble House
Unity Temple
Frederick C. Robie House
Postsparkasse
Schindler House
Fiat Works
Schröderhuis
Lovell Beach House
Sagrada Família
Stockholm Public Library
German Pavilion (Barcelona Pavilion)
Villa Savoye
Paimio Sanatorium
La Maison de Verre
Edgar J. Kaufmann House (Fallingwater)
Casa del Fascio
S.C. Johnson & Son Headquarters
Villa Mairea
The Woodland Cemetery
Casa Malaparte
Glass House
Golconde, Pondicherry
Barragán House and Studio
The Eames House (Case Study House No. 8)
Farnsworth House
L'Unité d'Habitation
Lever House
Kresge College
Chapelle Notre-Dame du Haut
Sydney Opera House
Crown Hall
Mill Owners' Association Building
St. Mark's Church in Björkhagen
Seagram Building
National Congress, Brazil
Louisiana Museum of Modern Art
Solomon R. Guggenheim Museum
Capilla de las Capuchinas
Couvent Sainte-Marie de La Tourette
Municipal Orphanage
TWA Flight Center
Ena de Silva House
The Beinecke Rare Book & Manuscript Library
Berliner Philharmonie
Sabarmati Ashram
Leicester University Engineering Building
National Museum of Anthropology
National Olympic Gymnasium, Tokyo
Rudolph Hall
Vanna Venturi House
Salk Institute
The Assembly, Chandigarh
Sea Ranch Condominium Complex
Piscinas de Marés
Montessori School, Delft
Habitat '67
Ford Foundation Headquarters
Museu de Arte de São Paulo
FAU, University of São Paulo
Itamaraty Palace
Phillips Exeter Academy Library
Kimbell Art Museum
Castelvecchio Museum
Douglas House
Marie Short House
SESC Pompéia
Centre Pompidou
Brion Family Tomb
Gehry House
Hedmark Cathedral Museum
Warehouse, Port of Montevideo
Koshino House
National Assembly Building, Bangladesh
Kate Mantilini, Beverly Hills
Hongkong and Shanghai Bank Headquarters
National Museum of Roman Art
Ricola Storage Building
Church of the Light
Institut du Monde Arabe
Museum of Sculpture, São Paulo
Kiasma
Igualada Cemetery
Parc de la Villette
Vitra Fire Station
Maison à Bordeaux
Neurosciences Institute, La Jolla
Therme Vals
Guggenheim Museum Bilbao
Kunsthaus Bregenz
Signal Box Auf dem Wolf
Chapel of St. Ignatius
Diamond Ranch High School
Jewish Museum Berlin
Boyd Art Center, Riversdale
Sendai Mediatheque
American Folk Art Museum

英汉对照表与注释

Ana Tostões　安娜·托斯通伊斯
Andrea Simitch + Val K. Warke　安德烈娅·西米奇 + 瓦尔·K·沃克
Antoine Predock　安托万·普雷多克
Ben van Berkel + Caroline Bos / UNStudio　本·范贝克尔 + 卡罗琳·博斯 /UNS 工作室
Bernard Tschumi　伯纳德·屈米
Carme Pinós　卡梅·皮诺斯
César Pelli　西萨·佩里
Craig Hodgetts　克雷格·霍杰茨
Dame Zaha Hadid　扎哈·哈迪德女士
Daniel Libeskind / Studio Libeskind　丹尼尔·里伯斯金 / 里伯斯金工作室
Dominique Perrault　多米尼克·佩罗
Elizabeth Diller + Ricardo Scofidio + Charles Renfro　伊丽莎白·迪勒 + 里卡多·斯科菲迪奥 + 查尔斯·伦弗罗
Enrique Norten / TEN Arquitectos　恩里克·诺尔滕 /TEN 事务所
Eric Owen Moss　埃里克·欧文·莫斯
Francine Houben　弗朗辛·乌邦
Fumihiko Maki　槇文彦
Greg Lynn　格雷格·林恩
Henry N. Cobb　亨利·N·科布
Hernan Diaz Alonso　埃尔南·迪亚斯·阿隆索
Hitoshi Abe　阿部仁史
Jeanne Gang / Studio Gang　珍妮·甘 /Gang 工作室
James S. Polshek　詹姆斯·S·波尔舍克
Jesse Reiser + Nanako Umemoto　杰西·赖泽 + 梅本奈奈子
Jong Soung Kimm　金钟成
Jorge Silvetti　豪尔郝·西尔韦蒂
José Oubrerie　若泽·乌贝里
Kazuyo Sejima + Ryue Nishizawa / SANAA　妹岛和世 + 西泽立卫 /SANNA 事务所
Kengo Kuma　隈研吾
Kevin Roche　凯文·罗奇
Lars Lerup　拉斯·勒普
Léon Krier　莱昂·克里尔
Mack Scogin　麦克·斯科金
Marion Weiss + Michael Manfredi　马里昂·韦斯 + 迈克尔·曼弗雷迪
Marlon Blackwell　马龙·布莱克韦尔
Michael Maltzan　米夏埃尔·马尔灿
Mónica Ponce de León　莫妮卡·庞塞·德莱昂
Moshe Safdie　摩西·萨夫迪
Neil Denari　尼尔·德纳里
Odile Decq　奥迪勒·德克
Peter Eisenman　彼得·埃森曼
Preston Scott Cohen　普雷斯顿·斯科特·科恩
Rafael Moneo　拉斐尔·莫内欧
Richard Meier　理查德·迈耶
Richard Rogers　理查德·罗杰斯
Robert A.M. Stern　罗伯特·A·M·斯特恩

Robert Venturi + Denise Scott Brown　罗伯特·文丘里 + 丹尼斯·斯科特·布朗
Shigeru Ban　坂茂
Sir Peter Cook　彼得·库克爵士
Stan Allen　斯坦·艾伦
Stanley Saitowitz　斯坦利·塞陶维兹
Steven Holl　斯蒂文·霍尔
Tadao Ando　安藤忠雄
Thom Mayne / Morphosis　汤姆·梅恩 / 墨菲西斯事务所
Tod Williams + Billie Tsien　托德·威廉姆斯 + 比利·钱
Toyo Ito　伊东丰雄
Winka Dubbeldam　文卡·度别丹
Winy Maas + Jacob van Rijs + Nathalie de Vries　维尼·马斯 + 雅各布·范里斯 + 娜塔莉·德弗里斯
Wolf D. Prix / COOP HIMMELB(L)AU　沃尔夫·D·普瑞 / 蓝天组

AEG Turbine Factory　AEG 透平机车间
Bagsværd Church　鲍斯韦教堂
Baker House　贝克公寓
Barragán House and Studio　巴拉甘自宅与工作室
Bauhaus Dessau　包豪斯德绍校舍
Berliner Philharmonie　柏林乐团音乐厅
Beurs van Berlage　贝尔拉赫证券交易所
Brion Family Tomb　布里翁墓园
Carpenter Center for the Visual Arts　卡朋特视觉艺术中心
Casa del Fascio　法西斯宫
Casa Malaparte　马拉帕特别墅
Casa Milà　米拉公寓
Castelvecchio Museum　古堡博物馆
Centraal Beheer Building　中央贝赫保险公司大楼
Centre Le Corbusier (Heidi Weber House)　柯布西耶中心
Centre Pompidou　蓬皮杜中心
Chapelle Notre-Dame du Haut　朗香教堂
Couvent Sainte-Marie de La Tourette　拉图雷特修道院
Crown Hall　克朗楼
Diamond Ranch High School　钻石农场中学
Dymaxion House　动态高效住宅
Edgar J. Kaufmann House (Fallingwater)　流水别墅
Einstein Tower　爱因斯坦天文台
Farnsworth House　范斯沃斯住宅
Fiat Works　菲亚特工厂
Ford Foundation Headquarters　福特基金会总部大楼
Free University of Berlin　柏林自由大学
Frederick C. Robie House　罗比住宅
Gallaratese II Apartments　加拉拉特西公寓
Gamble House　甘布尔住宅
Gehry House　盖里自宅
German Pavilion (Barcelona Pavilion)　德国馆

Glasgow School of Art　格拉斯哥艺术学校
Glass House　玻璃屋
Guggenheim Museum Bilbao　毕尔巴鄂古根海姆博物馆
Habitat '67　栖息地67号
Hillside Terrace Complex I – VI　代官山集合住宅
Hongkong and Shanghai Bank Headquarters　香港汇丰银行大厦
House VI　六号住宅
Igualada Cemetery　伊瓜拉达墓园
Institut du Monde Arabe　阿拉伯世界研究中心
Jewish Museum Berlin　柏林犹太人博物馆
Kimbell Art Museum　金贝尔美术馆
La Maison de Verre　玻璃之家
Larkin Company Administration Building　拉金公司行政大楼
Leicester University Engineering Building　莱斯特大学工程楼
Lever House　利华大厦
Lloyd's of London　伦敦劳埃德大厦
Lovell Beach House　洛弗尔海滨住宅
Lovell Health House　洛弗尔健康之家
L'Unité d'Habitation　马赛公寓
Maison à Bordeaux　波尔多住宅
Michaelerplatz House (Looshaus)　米歇尔广场公寓
Mill Owners' Association Building　棉纺织协会总部
Municipal Orphanage, Amsterdam　阿姆斯特丹市孤儿院
Museu de Arte de São Paulo　圣保罗艺术博物馆
Nakagin Capsule Tower　中银舱体大楼
National Assembly Building, Bangladesh　孟加拉国国会大厦
National Congress, Brazil　巴西议会大厦
National Museum of Roman Art　西班牙国家罗马艺术博物馆
National Olympic Gymnasium　代代木国立综合体育馆
Neue Staatsgalerie　斯图加特美术馆
New National Gallery　国家美术馆新馆
Parc de la Villette　拉维莱特公园
Phillips Exeter Academy Library　菲利普斯埃克塞特中学图书馆
Piscinas de Marés　莱萨浴场
Postsparkasse　邮政储蓄银行
Olympic Stadium, Munich　慕尼黑奥林匹克体育场
Querini Stampalia Renovation　奎里尼·斯坦帕利亚基金会更新
Rooftop Remodeling Falkestrasse　屋顶加建
Rudolph Hall (Yale Art and Architecture Building)　鲁道夫馆
Rusakov Workers' Club　鲁萨科夫工人俱乐部
Salk Institute　萨尔克生物研究所
Säynätsalo Town Hall　赛于奈察洛市政厅
Schindler House　申德勒自宅
S.C. Johnson & Son Headquarters　约翰逊制蜡公司总部
Schröderhuis　施罗德住宅
Seagram Building　西格拉姆大厦
Sendai Mediatheque　仙台媒体中心
SESC Pompéia　庞培娅艺术中心
Smith House　史密斯住宅
Solomon R. Guggenheim Museum　所罗门古根海姆博物馆

Stahl House (Case Study House No. 22)　斯塔尔住宅
Stockholm Public Library　斯德哥尔摩公共图书馆
Sydney Opera House　悉尼歌剧院
The Assembly, Chandigarh　昌迪加尔议会大厦
The Beinecke Rare Book & Manuscript Library　拜内克古籍善本图书馆
The Breuer Building (Whitney Museum of American Art)　布罗伊尔楼
The Eames House (Case Study House No. 8)　埃姆斯自宅
The Woodland Cemetery　林地墓园
Tugendhat House　图根哈特住宅
Turin Exhibition Hall　都林展览馆
TWA Flight Center　环球航空公司候机楼
Van Nelle Factory　范内勒工厂
Vanna Venturi House　母亲住宅
Villa Mairea　玛丽亚别墅
Villa Müller　穆勒别墅
Villa Savoye　萨伏伊别墅
Vitra Fire Station　维特拉消防站
Yokohama International Port Terminal　横滨国际港客运中心

本书注释均为译者注。

05

1. 原文为 S.C.Johnson& Son,其官方名称是 S.C.庄臣父子公司。在涉及其总部大楼时,国内一般译为"约翰逊公司总部"或"约翰逊制蜡公司总部"。
2. 派莱克斯玻璃管(The Pyrex tubes),一种由耐热玻璃制成的实验试管。

07

1. 康按照路易斯·巴拉甘(Luis Barragan)的建议设计了空敞的中心庭院。原本的庭院设计方案是在水渠两边种植意大利柏树。康邀请巴拉甘来到现场后,巴拉甘说"一片叶子也不要放进来",同时表示这个庭院是"将两侧建筑联系起来的、朝向天空的立面"。

08

1. 装饰派艺术(Art Deco),起源于 1920 年前后,具有靓丽鲜艳的色彩和明朗粗犷的轮廓、多采用几何和流线型,是现代主义早期的一种装饰形式。
2. 原文为 wide-flange columns,指依靠法兰方式连接(先将结构或部件固定在法兰盘上,两个法兰盘之间再加入法兰垫,然后用螺栓连接在一起)的宽柱。

09

1. 多明我会(Dominican Order),也称为"多米尼克教派"。天主教托钵修会主要派别之一。1217 年由西班牙人多明戈创立。该教派认为"光是神圣知识的重要媒介",这一点也影响了拉图雷特修道院的设计。——译者注
2. 伊安尼斯·克塞纳基斯(Iannis Xenakis), 20 世纪激进作曲家,其革命性观念包括对音乐组织的系统化和数学性理解,以及他将音乐结构与建筑所做的比照。

10

1. 环球航空公司候机楼依靠建筑结构本体的非凡形态达成形式表现,而盖里设计的毕尔巴鄂古根海姆博物馆则主要借助参数化设计的建筑表皮展现曲折而动态的建筑外形。

13

1. 流水别墅(FALLINGWATER)的正式名称是艾德加·J·考夫曼住宅(EDGAR J. KAUFMANN HOUSE),由于"流水别墅"的名称太过精彩,国内外一般都直接使用这一名称指代该建筑。

16

1. 截头锥体(frustum),即俗称的"漏斗形"截去尖头后得出的形状。

17

1. 埃姆斯自宅也被称为 8 号案例住宅。

19

法尔尼斯府邸（Palazzo Farnese），是罗马一座杰出的文艺复兴建筑，由小桑迦洛（Antonio da Sangallo the Younger）、米开朗基罗（Michelangelo Buonarroti）等人设计。

20

二战之后，柯布西耶曾设计过一系列城市集合住宅，并称为"居住单位"（L'Unité d'Habitation）。本文所述的居住单位位于法国马赛，长久以来一直被译为"马赛公寓"。为了方便理解，本书沿用了这一译法。

22

1　学院派艺术（Beaux-Arts），也称为"布杂体系"。指起源于巴黎美术学院的艺术与建筑教育方式，着重建筑形式的训练，对学生的美术功底要求较高。

23

1　风格派（De Stijl），1917 年由蒙德里安等人在荷兰创立。其宗旨是创作时不使用任何具象元素，只用颜色和几何形象来表现纯粹的精神。

2　原文为"Dematerializes"，意为祛除建筑的具象功能特征（如传统的窗台、檐口、门廊等要素），而着重反映线条、平面等抽象元素的组织关系与意义。

24

1　新陈代谢派（Metabolism）是日本在 1960 年前后形成的建筑创作组织，受丹下健三的影响，强调事物的生长、变化与衰亡，极力主张采用新的技术来解决建筑和城市问题，代表人物是大高正人、菊竹清训和黑川纪章。

27

1　根据阿尔托大学裘振宇博士的考证研究，伍重计划在歌剧院的巨型外部结构之下，再设一个巨型内部结构作为顶棚，同时处理室内空间和声学问题（正如他其后在鲍斯韦教堂所做的设计）。伍重已经做出了一部分图纸与结构模型，但并没有被官方采纳。

28

1　皮拉内西 (GiovanniBattistaPiranesi)，18 世纪意大利雕刻家、建筑师。其作品的特征是强烈的光影和空间对比，以及对细节的准确描绘。

30

1　透平机（turbine），即涡轮机。

33

1　卢西奥·科斯塔（Lúcio Costa）是 20 世纪巴西著名规划家，也是一位现代主义建筑师，代表作品是 1957 年的巴西利亚规划。他对记录和保护巴西建筑遗产作出了巨大贡献。

39

1 克劳德-尼古拉斯·勒杜（Claude-Nicolas Ledoux），18世纪法国建筑师、城市规划师，也是新古典主义建筑的标志性人物。

42

1 瓦西里·康定斯基（Wassily Kandinsky），生于俄罗斯，是20世纪欧洲著名的画家和美术理论家，也是欧洲抽象艺术的先驱。

2 翁贝特·博乔尼（Umberto Boccioni），是意大利画家和雕塑家，是未来主义画派的发起者和核心人物。

45

1 玻璃屋和范斯沃斯住宅的墙体都是透明的玻璃，在设计之初也没有考虑安装窗帘，只能依靠业主广阔的私人土地使住宅变得私密。

48

1 罗伯特·劳申贝格（Robert Rauschenberg），美国现代艺术家，擅长将多种材质拼贴组合，创造了"综合绘画"（combines-painting）的表现技巧。

2 马塞尔·杜尚（Marcel Duchamp），生于法国，1954年移居美国，是二战以后最重要的西方现代艺术家之一。下文中的"现成品艺术"是指将批量生产的物品重新组装布置，使其获得新的审美价值，杜尚的代表作品"泉"即属于这一类作品。

51

1 路易吉·莫雷蒂（luigi moretti），意大利建筑师，活跃于20世纪30年代。他注重建筑真实结构的表现，善于塑造建筑形体和材质。

2 埃德温·勒琴斯（Edwin Lutyens），英国建筑师，活跃于20世纪早期。勒琴斯曾设计过很多英国田园别墅，最著名的作品是新德里总督府（Rashtrapati Bhavan）。

3 查尔斯·福林·麦金（Charles Follen McKim），美国学院派建筑师，活跃于19世纪末20世纪初。

53

1 住宅顶层的一部分使用了乳白色的弧线转角玻璃。

54

1 马萨诸塞州（Massachusetts）位于美国东北部，属于温带气候且四季分明，夏季短，冬季长。

55

1 该建筑全封闭的大理石表面是为了隔绝强烈光线和气流扰动对珍本书籍带来的损害，建筑底层架空则是为了防止内部进水，可以说，建筑采取"与世隔绝的姿态"不仅是形式上的设计，更是基于珍本图书储藏的功能性需求。

2 野口勇（Isamu Noguchi），日裔美国人，是20世纪最著名的雕塑家之一，也是最早尝试将雕塑和景观设计结合的人。

58

1　分形，通常定义为"一个零碎的几何形状，可以分为数个部分，且每一部分都（至少近似地）是整体缩小后的形状"，即具有自相似的性质。

59

1　按照路易斯·康的"服务空间"与"被服务空间"观念，封闭围合的服务空间（即辅助功能空间，如楼梯间、卫生间甚至办公室）本身就可以起到支撑整体结构的作用，也就是本文中的"带有功能的柱子"。

60

1　柯林·罗 (Colin Rowe)，著名建筑和城市历史学家、批评家和理论家。主要著作是《拼贴城市》。他主张现代建筑应摒弃纯粹的抽象，自然地随历史的演进而自然演化。斯特林曾是柯林·罗的学生，其作品也受到柯林·罗理论的影响。

65

1　Gargoyles，也称为石像鬼，类似中国古建筑中的脊兽。常以长有翅膀的怪兽形象出现，有守护建筑的寓意。一般布置在教堂等大型建筑的屋檐处，常起到排水口的作用。

66

1　受到气候因素的影响，传统的阿拉伯建筑通常将室外活动空间包围在建筑内部，由墙体或房间围合，组织成室内庭院的形式。

68

1　三条轴线的主题分别是大屠杀之轴（axis of the holocaust）、逃亡之轴（axis of exile）和延续之轴（axis of continuity），大屠杀之轴通向屠杀之塔，逃亡之轴通向逃亡者之园，延续之轴则将人们引入建筑上层的其他展示空间。

72

1　这座古城堡始最早可追溯至罗马时期，1801年之前便形成了"三面由城墙围合，北侧向阿迪济河（Adige river）开敞"的格局。1802年开始，法国拿破仑军队在古堡北侧沿阿迪济河修建了城墙和两层高的营房（如今的古堡博物馆主体便是这栋营房）。斯卡帕重视历史层次的表达，为了展示12世纪的城门与14世纪的城壕，拆除了北侧营房的一间。

74

1　菲舍尔·冯·埃拉赫（Johann Bernhard Fischer von Erlach），是一位17世纪奥地利建筑家、雕塑家和建筑史学家。他创作的巴洛克风格曾深刻地影响并塑造了哈布斯堡帝国的建筑品味。

77

1　1945年，美国《艺术与建筑》（Arts & Architecture）杂志的主编约翰·恩腾扎发起了"案例住宅计划"（Case Study House program），邀请当时的先锋建筑师设计样板住宅并试图向普通民众推广，产生了一系列住宅设计作品。本书中的埃姆斯自宅、斯塔尔住宅都从属于这个计划。
2　美国建筑摄影家朱利叶斯·舒尔曼（Julius Shulman）曾为斯塔尔住宅拍摄过一张夜景照片，在当时产生了巨大影响，斯塔尔住宅也因此声名远播。

79

1　毕德麦雅（Biedermeier），指德意志邦联诸国在1815年（《维也纳公约》签订）至1848年（资产阶级革命开始）之间的历史时期，现则多指文化史上的中产阶级艺术时期。

85

1　奎里尼府邸始建于1513年，建筑主体长期遭受洪水损害——每到冬季，威尼斯潟湖的海水灌入建筑底层，结构体被逐渐侵蚀，内庭院也处于废弃状态。
2　斯卡帕作品中多次出现的马赛克装饰是与其好友、威尼斯画家马瑞欧·德路易吉（Mario De luigi）合作的产物。

86

1　原文为"the second wave of postmodernity"。按照作者的意思，第一次浪潮是"后现代形式主义"（Postmodern-formalism），即吸收各种历史元素的折中风格，第二次浪潮即埃森曼等人发起的"解构主义建筑"风格（Deconstruction Architecture）。

88

1　总体艺术，即"gesamtkunstwerk"或"total work of art"，指融合了多种常规艺术门类的艺术表现方式。例如歌剧便是音乐、戏剧和视觉艺术综合后的结果。

91

1　"第十次小组"，也称为"小组十"（Team X or Team Ten）。由CIAM（国际现代建筑师协会）第九次会议（1953年）中的积极分子——英国的史密森夫妇（Alison and Peter Smithson）与荷兰的阿尔多·范艾克等人发起，强调城市与建筑的情感需求——社区感、归属感、邻里感与场所感。他们批评并挑战了由老一辈建筑师树立的教条主义的都市生活方式（如《雅典宪章》中提出的居住、工作、游憩与交通四大功能），并最终造成CIAM的分裂与最终休会（1956年）。
2　毯式建筑（Mat-building）是由艾莉森·史密森（Alison Smithson）提出的建筑术语，指一种大尺度高密度的建筑类型，以精确调节的平面网格为基础，总体与局部中一般包含着特定的尺度模数。建筑高度一般不高，占地面积较大，各单元之间穿插布置着庭院或小广场。代表建筑作品有阿姆斯特丹市孤儿院、柏林自由大学等。

92

1　《勒·柯布西耶全集》（Le Corbusier: Oeuvre Complète）是柯布西耶的建筑作品集，按照作品的完成时间共列为8卷，收录了其整个职业生涯(1910年至1969年)所做的绝大部分建筑

93

1. 谢德拉克·伍兹（Shadrach Woods）美国建筑师、规划师、同时也是一位建筑理论家。早年曾在勒·柯布西耶的事务所工作，之后与"第十次小组"建立了紧密关系，其代表性的观点是"干茎"（stem）与"网络"（web）理论。

95

1. 在两座体育馆塔楼施工时，采用了首先浇筑井字梁楼板，然后在楼板之上浇筑围墙的特殊方式，因此在建筑立面可以看到每一层楼板的施工痕迹与楼层界线，形成了独特的表皮肌理。从水塔的立面可以推断，文中提到的"外渗浇筑"技术应是类似的分层浇筑方法。

97

1. 卡洛·艾莫尼诺（Carlo Aymonino，1926—2010 年），意大利建筑师、城市规划师。最为著名的作品便是与罗西合作的米兰市加拉拉特西住宅综合体，这座住宅项目的 A1、A2、B 及 C 栋由艾莫尼诺设计，而 D 栋则由罗西设计。
2. 赛罗塔石碑（Rosetta Stone），发现于 1799 年的一块刻有埃及象形文的石碑，现在人们正是通过它逐步破解了埃及象形文字。
3. 原文为"'degree zero' of architectural forms"，罗西的类型学与此概念密切相关。简单地说就是抛却现有建筑的一些模式、程式，将空间、建筑最初的原型作为设计的起始点。

99

1. 扎哈·哈迪德于 1979 年开设了自己的设计事务所，33 岁时在香港山顶俱乐部国际竞标中一鸣惊人拔得头筹，但是由于开发商失去了场地所有权而没有建造。整个二十世纪八十年代，由于各种原因哈迪德的设计方案都停留在图纸阶段，1993 年的维特拉消防站是扎哈·哈迪德的第一项建成作品。

参考文献

萨伏伊别墅　VILLA SAVOYE

1/ Benton, Tim.The Villas of Le Corbusier, 1920 – 1930: With Photographs in the Lucien Hervé Collection. New Haven: Yale University Press, 1987. 2/ Corbusier, Le.Towards a New Architecture. New York: Dover Publications, 1986. 3/ Curtis, William J. Le Corbusier, Ideas and Forms. Oxford: Phaidon, 1986. 4/ Davies, Colin.Key Houses of the Twentieth Century: Plans, Sections and Elevations. New York: W.W. Norton, 2006. 5/ Corbusier, Le, Pierre Jeanneret, Willy Boesiger, Oscar Stonorov, and Max Bill.OEuvre Complète. Basel: Birkhäuser, 1999. 6/ Frampton, Kenneth, and Yukio Futagawa.Modern Architecture1920 – 1945. New York: Rizzoli, 1983.

朗香教堂　CHAPELLE NOTRE-DAME DU HAUT

1/ Corbusier, Le.The Chapel at Ronchamp. United Kingdom: Architectural Press London, 1957. 2/ Crippa, Maria Antonietta, Françoise Caussé, and Caroline Beamish. Le Corbusier: The Chapel of Notre-Dame Du Haut at Ronchamp. London: Royal Academy of Arts, 2015. 3/ Pauly, Danièle. Le Corbusier: The Chapel at Ronchamp. Paris: Fondation Le Corbusier, 2008. 4/ Curtis, William J. R. Le Corbusier: Ideas and Forms. Oxford: Phaidon, 1986. 5/ Palmes, J. C. and Maurice Jardot.Le Corbusier: Creation Is a Patient Search. New York: Praeger, 1960.

巴塞罗那博览会德国馆　GERMAN PAVILION (BARCELONA PAVILION)

1/ Neumeyer, Fritz.Global Architecture. Japan: A.D.A. EDITA Tokyo Co. Ltd, 1995. 2/ Solà-Morales, Ignasi De, Cristian Cirici, and Fernando Ramos. Mies Van Der Rohe: Barcelona Pavilion. Barcelona: Gili, 1993. 3/ Dodds, George.Building Desire: On the Barcelona Pavilion. London: Routledge, 2005. 4/ Frampton, Kenneth, and Yukio Futagawa.Modern Architecture 1920 – 1945. New York: Rizzoli, 1983. 5/ Johnson, Philip.Mies Van Der Rohe. New York: Museum of Modern Art, 1978. 6/ Fritz Neumeyer.Mies Van Der Rohe: German Pavilion, International Exposition, Barcelona, Spain, 1928 – 29 (reconstructed 1986), Tugendhat House, Brno, Czechoslovakia, 1928 – 30. Tokyo: A.D.A. EDITA Tokyo, 1995. 7/ Evans, Robin, "Mies van der Rohe's Paradoxical Symmetries," in Translations from Drawing to Building and Other Essays. 8/ Quetglas, Josep, Fear of Glass, Birkhauser, 2001.

蓬皮杜中心　CENTRE POMPIDOU

1/ Piano, Renzo.Du plateau Beaubourg au Centre Georges Pompidou. Paris: Éditions du Centre Pompidou, 1987. 2/ Rattenbury, Kester.Richard Rogers: The Pompidou Center. England: Oxon; New York: Routledge, 2012.
3/ Futagawa, Yukio.Global Architecture 44: Piano + Rogers Architects ONE ARUP Engineers. Tokyo, A.D.A. EDITA, 1975.

4/ Weston, Richard.Key Buildings of the 20th Century: Plans, Sections and Elevations. New York: W.W. Norton & Co., 2010.

约翰逊制蜡公司总部　S.C. JOHNSON & SON HEADQUARTERS

1/ Hertzberg, Mark. Frank Lloyd Wright's SC Johnson Research Tower. San Francisco: Pomegranate, 2010. 2/ Carter, Brian. Johnson Wax Administration Building and Research Tower. London: Phaidon Press, 1998. 3/ Wright, Frank Lloyd.Johnson & Son, Administration Building and Research Tower. Tokyo: A.D.A. EDITA Tokyo, 1970, 1974 printing. 4/ Weston, Richard. Key buildings of the 20th Century. New York: W.W. Norton & Co., 2010. 5/ Lipman, Jonathan.Frank Lloyd Wright &The Johnson Wax Building. NY: Rizzoli, 1986.

范斯沃斯住宅　FARNSWORTH HOUSE

1/ Goldberger, Paul; Phyllis Lambert, and Sylvia Lavin.Modern Views: Inspired by the Mies Van Der Rohe Farnsworth House and the Philip Johnson Glass House. New York, NY, USA: Assouline Pub., 2010. 2/ Jenna Cellini, Elizabeth Milnarik, Brad Roeder.Farnsworth House Recording Project. United States: National Trust For Historic Preservation, 2009. 3/ Rohe, Ludwig Mies Van Der, and Dirk Lohan.Mies Van Der Rohe: Farnsworth House, Plano, Illinois, 1945 - 50. Tokyo, Japan: A.D.A. EDITA Tokyo, 2000. 4/ Clemence, Paul.Mies Van Der Rohe's Farnsworth House. Atglen, PA: Schiffer Pub., 2006.
5/ Vandenberg, Maritz.Farnsworth House: Ludwig Mies Van Der Rohe. London: Phaidon, 2003. 6/ Wagner, George, "The Lair of the Bachelor," in Architecture and Feminism, ed. Coleman, Danze, and Henderson. Princeton: Princeton Architectural Press, 1997.

萨尔克生物研究所　SALK INSTITUTE

1/ Stoller, Ezra.The Salk Institute. New York: Princeton Architectural Press, 1999. 2/ James Steele.Salk Institute. London: Phaidon, 1993. 3/ Savio, Andrea, and Louis Isidore. Louis I. Kahn: Salk Institute. Firenze: Alinea, 1994.
4/ Goldhagen, Sarah Williams.Louis Kahn's Situated Modernism. New Haven, CT: Yale University Press, 2001.

玻璃之家　LA MAISON DE VERRE

1/ Bauchet, Bernard, and Marc Vellay, with Yukio Futagawa (ed.). La Maison de Verre. Tokyo: A.D.A. EDITA, 1988. 2/ Cinqualbre, Olivier.Pierre Chareau, la Maison de verre, 1928 - 1933: un objet singulier. Paris: J.-M. Place, 2001.
3/ Frampton, Kenneth, and Yukio Futagawa.Modern Architecture 1920 - 1945. New York: Rizzoli, 1983. 4/ Chareau, Pierre.Pierre Chareau: Architecte, Un Art Interieur. Paris: Centre Georges Pompidou, 1993. 5/ Taylor, Brian Brace.Pierre Chareau: Designer and Architect. Köln: Benedikt Taschen, 1992. 6/ Vellay, Marc, and Kenneth Frampton.Pierre Chareau:

Architect and Craftsman, 1883‑1950. New York: Rizzoli, 1985.

拉图雷特修道院　COUVENT SAINTE-MARIE DE LA TOURETTE

1/ Petit, Jean.Un couvent de Le Corbusier: Volume réalisé. Paris: Cahiers Forces Vives-Editec, 1961. 2/ Pirazzoli, Giacomo. Le Corbusier a La Tourette:qualche congettura. Firenze: All'insegna del giglio, 2000. 3/ Potié, Philippe. Le Corbusier: le Couvent Sainte Marie de La Tourette = the Monastery of Sainte Marie de La Tourette. Paris: Fondation le Corbusier; Boston: Birkhäuser Publishers, 2001. 4/ Corbusier, Le; Yukio Futagawa, and Arata Isozaki.Couvent Sainte Marie De La Tourette, Eveux-sur-l'Arbresle, France 1957‑60. Tokyo: A.D.A. EDITA, 1971. 5/ Cohen, Jean-Louis.Le Corbusier: Le Grand. London: Phaidon, 2008. 6/ Hervé, Lucien, The Architecture of Truth: The Cistercian Abbey of Le Thoronet.Introduction by Le Corbusier. London: Thames and Hudson, 1957. 7/ Rowe, Colin, "La Tourette," in The Mathematics of the Ideal Villa and Other Essays. Cambridge: The MIT Press, 1976.

环球航空公司候机楼　TWA FLIGHT CENTER

1/ Saarinen, Eero.TWA Terminal Building, Kennedy Airport, New York, 1956‑62: Dulles International Airport (Washington, D.C.). Tokyo: A.D.A. EDITA Tokyo, 1973. 2/ Stoller, Ezra.The TWA Terminal. New York: Princeton Architectural Press, 1999. 3/ Weston, Richard.Key Buildings of the 20th Century: Plans, Sections, and Elevations. New York: W.W. Norton & Company, 2010.
4/ Ringli, Kornel.Designing TWA: Eero Saarinen's Terminal in New York. Zürich: Park, 2015.

柏林爱乐音乐厅　BERLINER PHILHARMONIE

1/ Scharoun, Hans, Yukio Futagawa, and Hiroshi Sasaki.The Berlin Philharmonic Concert Hall, Berlin, West Germany, 1956, 1960‑63. Tokyo: A.D.A. EDITA Tokyo, 1973. 2/ Barnbeck, Ulla.Architekten, Hans Scharoun. Stuttgart: IRB Verlag, 1987. 3/ Weston, Richard.Key Buildings of the 20th Century: Plans, Sections and Elevations. New York: W.W. Norton, 2010. 4/ Forsyth, Michael.Buildings for Music. Cambridge: MIT Press, 1985.

毕尔巴鄂古根海姆博物馆　GUGGENHEIM MUSEUM BILBAO

1/ Museo Guggenheim (Bilbao).Guggenheim Museum Bilbao Collection. Madrid: TF Editores, 2009. 2/ Gehry, Frank O. GA Document; 54: Guggenheim Bilbao Museoa. New York: GA Intl Company Limited, 1998. 3/ Bruggen, Coosje Van. Frank O. Gehry: Guggenheim Museum Bilbao. New York, NY: Guggenheim Museum Publications, 1998. 4/ Gilbert-Rolfe, Jeremy, and Frank O. Gehry.Frank Gehry: The City and Music. London: Routledge, 2002. 5/ Isenberg, Barbara, and Frank O. Gehry.Conversations with Frank Gehry. New York: Alfred A.

Knopf, 2009. 6/ Gehry, Frank O., and Germano Celant.Frank O. Gehry. N.p.: n.p., n.d. 7/ Gehry, Frank O., Fernando Márquez Cecilia, and Richard C. Levene. Frank Gehry, 1987–2003. Madrid: El Croquis Editorial, 2006. 8/ Mathewson, Casey C. M., and Frank O. Gehry.Frank O. Gehry: Selected Works: 1969 to Today. Richmond Hill, Ont.: Firefly, 2007.

流水别墅　EDGAR J. KAUFMANN HOUSE (FALLINGWATER)

1/ Wright, Frank Lloyd.Kaufmann House, Fallingwater, Bear Run Pennsylvania, 1936. Tokyo: A.D.A. EDITA Tokyo, 1970. 2/ Davis, Colin.Key Houses of the Twentieth Century. New York: W.W. Norton, 2006. 3/ Kaufmann, Edgar.Fallingwater, a Frank Lloyd Wright Country House. New York: Abbeville, 1986. 4/ Frampton, Kenneth, and Yukio Futagawa.Modern Architecture 1920–1945. New York: Rizzoli, 1983.

林地墓园　THE WOODLAND CEMETERY

1/ Asplund, Erik Gunnar; Dan Cruickshank, and Martin Charles. Erik Gunnar Asplund. London: Architects Journal, 1988. 2/ Caldenby, C. Asplund. Stockholm: Arkitektur Förl., 1986. Print. 3/ Caldenby, Claes, and Olof Hultin.Asplund. New York: Rizzoli, 1986. 4/ Lewerentz, Sigurd; Claes Dymling, Wilfried Wang, and Fabio Galli.Architect Sigurd Lewerentz. Stockholm: Byggförlaget, 1997. 5/ Frampton, Kenneth, and Yukio Futagawa. Modern Architecture,1920–1945. New York: Rizzoli, 1983. 6/ Holmdahl, Gustav; Sven Ivar Lind, and Kjell Ödeen.Gunnar Asplund Architect, 1885–1940. Stockholm: AB Tidskriften Byggmästaren, 1950. 7/ Wrede, Stuart.The Architecture of Erik Gunnar Asplund. Cambridge: The MIT Press, 1980. 8/ Jones, Peter Blundell. Gunnar Asplund. New York: Phaidon Press Inc., 2006. 9/ Ahlin, Janne.Sigurd Lewerentz, Architect; 1885–1975. Cambridge: The MIT Press, 1987.

西格拉姆大厦　SEAGRAM BUILDING

1/ Weston, Richard. Key Buildings of the 20th Century: Plans, Sections and Elevations. New York: W.W. Norton, 2010. 2/ Zimmerman, Claire. Mies Van Der Rohe, 1886–1969: The Structure of Space. Hong Kong: Taschen, 2006. 3/ Gössel, Peter, and Gabriele Leuthäuser. Architecture in the Twentieth Century. Köln: Benedikt Taschen, 1991. 4/ Lambert, Phyllis, and Barry Bergdoll.Building Seagram. London: Yale University Press, 2013.

所罗门古根海姆博物馆　SOLOMON R. GUGGENHEIM MUSEUM

1/ Muto, Akira, and Yukio Futagawa.Global Architecture 12: Frank Lloyd Wright 1887–1959. Tokyo, A.D.A. EDITA, 1975. 2/ Pfeiffer, Bruce Brooks.Frank Lloyd Wright. Complete Works. Vol. 3, 1943–1959. Hong Kong; Los Angeles, Taschen, 2009.3/ Pfeiffer, Bruce Brooks.Frank Lloyd Wright, the Guggenheim Correspondence. Fresno Press at California State University,

1986.
4/ Ballon, Hilary.The Guggenheim: Frank Lloyd Wright and the Making of the Modern Museum. New York, London, Guggenheim Museum, 2009.
5/ Solomon R. Guggenheim Museum.The Solomon R. Guggenheim Museum. New York: Guggenheim Museum Publications, 1995.

埃姆斯自宅/8号案例住宅　THE EAMES HOUSE (CASE STUDY HOUSE NO. 8)

1/ Koenig, Gloria, and Peter Gössel. Eames. Köln: Taschen, 2015.
2/ Steele, James; Charles Eames, and Ray Eames.Eames House: Charles and Ray Eames. London: Phaidon, 2002. 3/ Davies, Colin.Key Houses of the Twentieth Century: Plans, Sections and Elevations. New York: W.W. Norton,2006. 4/ Gamboa, Pablo.La Casa Californiana Años 50. Bogota: Universidad Nacional De Colombia, 2007. 5/ Smith, Elizabeth A.T. Case Study Houses. Cologne: Taschen, 2007.

莱斯特大学工程楼　LEICESTER UNIVERSITY ENGINEERING BUILDING

1/ McKean, John.Leicester University Engineering Building: James Stirling and James Gowan. London: Phaidon, 1994. 2/ McKean, John.Pioneering British High-Tech: By Stirling & Gowan, Foster Associates and Richard Rogers Partnership. London: Phaidon Press, 1999. 3/ Yukio Futagawa.Leicester University Engineering Department, Leicester, Great Britain, 1959－63: Cambridge University History Faculty, Cambridge, Great Britain, 1964－66. Tokyo: A.D.A. EDITA, 1974. 4/ Berman, Alan.Jim Stirling and the Red Trilogy: Three Radical Buildings. London: Frances Lincoln, 2010. 5/ Maxwell, Robert.James Stirling/ Michel Wilford. Basel: Birkhäuser, 1998. 6/ Arnell, Peter, and Ted Bickford, eds. James Stirling: Buildings and Projects. New York: Rizzoli, 1984. 7/ Vidler, Anthony.James Frazer Stirling: Notes from the Archive. New Haven: Yale University Press, 2010.

法西斯宫　CASA DEL FASCIO

1/ Coppa, Alessandro.Guiseppe Terragni. Milan: 24 Ore Cultura, Pero, 2013. 2/ Terragni, Attilio Alberto; Daniel Libeskind, and Paolo Rosselli.The Terragni Atlas: Built Architectures. Milan: Skira, 2004. 3/ Eisenman, Peter, Giuseppe Terragni, and Manfredo Tafuri. Giuseppe Terragni: Transformations, Decompositions, Critiques. New York: Monacelli, 2003. 4/ Thiel-Siling, Sabine, and Wolfgang Bachmann.Icons of Architecture: The 20th Century. Munich: Prestel, 1998. 5/ Ching, Francis D. K.; Mark Jarzombek, and Vikramaditya Prakash.A Global History of Architecture. Hoboken, NJ: J. Wiley & Sons, 2007. 6/ Zevi, Bruno, Ommagio a Terragni. Milan: Etas/Kompass, 1968. 7/

Schumacher, Thomas L. Surface & Symbol: Giuseppe Terragni and the Architecture of Italian Rationalism. New York: Princeton Architectural Press, 1991.

马赛公寓　L'UNITÉ D'HABITATION
1/ Corbusier, Le; Pierre Jeanneret, Willy Boesiger, Oscar Stonorov, and Max Bill.OEuvre Complète. Basel: Birkhäuser, 1999. 2/ Corbusier, Le.Unité d'Habitation, Marseille-Michelet. Paris: Fondation Le Corbusier, 1983.
3/ Curtis, William J. R. Le Corbusier: Ideas and Forms. Oxford: Phaidon, 1986. 4/ Jenkins, David.Unite d'Habitation, Marseilles: Le Corbusier. London: Phaidon Press, 1993.

栖息地67号　HABITAT '67
1/ Kettle, John.Beyond Habitat. Cambridge, Massachusetts: The MIT Press, 1970. 2/ Safdie Moshe.Habitat 67.Montreal, Canada Queen's Printer, 1967.3/ Safdie Moshe and John Gray.Habitat; Moshe Safdie interviewed by John Gray. Montreal: Canada Tundra Books, 1967. 4/ Murphy, Diana and Moshe Safdie. Moshe Safdie. Mulgrave, Vic: 2009. 5/ Safdie Moshe and John Gray.Expo 67: Habitat. Montreal: Canada Tundra Books, 1967.

金贝尔美术馆　KIMBELL ART MUSEUM
1/ Vandenberg, Maritz, and Michael Brawne.Twentieth-Century Museums I: By Ludwig Mies van der Rohe, Louis Kahn and Richard Meier. London: Phaidon, 1999. 2/ Kahn, Louis I. Louis I. Kahn. Tokyo: A+u Pub. Co., 1974.3/ Kahn, Louis I.; David B. Brownlee, and David G. De Long.Louis I. Kahn: In the Realm of Architecture. New York, NY: Rizzoli, 2005. 4/ Lobell, John, and Louis I. Kahn.Between Silence and Light: Spirit in the Architecture of Louis I. Kahn. Boston: Shambhala, 2008.

施罗德住宅 SCHRÖDERHUIS　(RIETVELD SCHRÖDER HOUSE)
1/ Overy, Paul.The Rietveld Schroder House. Cambridge: MIT Press, 1988.
2/ Frampton, Kenneth, and Yukio Futagawa.Modern Architecture 1920 - 1945. New York: Rizzoli, 1983. 3/ Küper, Marijke; W. Quist, and Hans Ibelings. Gerrit Th. Rietveld: Casas Revista Internacional De Arquitectura (n° 39/40). Barcelona: Editorial Gustavo Gili, 2006. 4/ Mulder, Zijl, and Gerrit Thomas.Rietveld Schroder House. New York: Princeton Architectural Press, 1999.

中银舱体大楼　NAKAGIN CAPSULE TOWER
1/ Weston, Richard.Key Buildings of the 20th Century : Plans, Sections and Elevations. New York: W.W. Norton & Co., 2010. 2/ Ross, Michael Franklin.Beyond Metabolism: The New Japanese Architecture. New York: Architectural Record Books, 1978. 3/ Koolhaas, Charlie.Metaborizumu Torippu. Tokyo: Heibonsha, 2012. 4/ Schmal, Peter Cachola; Ingeborg

邮政储蓄银行　POSTSPARKASSE

1/ Hans, Hollein.Global Architecture. Japan: A.D.A. EDITA Tokyo Co. Ltd, 1978. 2/ Weston, Richard.Key Buildings of the 20th Century: Plans, Sections and Elevations. New York: W.W. Norton, 2010. 3/ Wagner, Otto.Otto Wagners Postsparkasse. Vienna: Zentralvereinigung der Architekten Österreichs, c.1975. 4/ Pile, John F. A History of Interior Design. United Kingdom: Laurence King, 2005. 5/ Geretsegger, Heinz, and Max Peintner.Otto Wagner 1841 – 1918: The Expanding City and the Beginning of Modern Architecture. New York: Rizzoli, 1979. 6/ Varnedoe, Kirk.Vienna 1900: Art, Architecture, & Design. New York: The Museum of Modern Art, 1986.

马拉帕特别墅　CASA MALAPARTE

1/ Davis, Colin.Key Houses of the Twentieth Century. New York: W.W. Norton, 2006. 2/ McDonough, Michael.Malaparte: A House Like Me. Clarkson Potter, 1999. 3/ Talamona, Marida. Casa Malaparte. New York: PrincetonArchitectural Press, 1992. 4/ Garofalo, Francesco, and Luca Veresane.Adalberto Libera. New York: Princeton Architectural Press, 1992. 5/ Curtis, William J. R. Modern Architecture Since 1900. London: Phaidon Press Limited 1996. 6/ Quilici, Vieri.Adalberto Libera: l'architettura come ideale. Rome: Officina Edizioni, 1981.

悉尼歌剧院　SYDNEY OPERA HOUSE

1/ Norberg, Schulz, Christian.Global Architecture. Japan: A.D.A. EDITA Tokyo Co. Ltd, 1980. 2/ Watson, Anne.Building a Masterpiece: The Sydney Opera House. Sydney: Powerhouse, 2006. 3/ Drew, Philip.Sydney Opera House: Jørn Utzon. London: Phaidon, 1995. 4/ Weston, Richard.Key Buildings of the 20th Century: Plans, Sections and Elevations. New York: W.W. Norton & Co., 2010. 5/ Møller, Henrik Sten, and Vibe Udsen. Jørn Utzon Houses. Copenhagen: Living Architecture Publishing, 2006.

鲁道夫馆　RUDOLPH HALL

1/ Monk, Tony.The Art and Architecture of Paul Rudolph. Chichester, West Sussex: Wiley-Academy, 1999. 2/ Rudolph, Paul, and Sibyl Moholy-Nagy.The Architecture of Paul Rudolph. New York: Praeger, 1970. 3/ Stoller, Ezra.The Yale Art + Architecture Building. New York: Princeton Architectural Press. 1999. 4/ Rudolph, Paul, and Yukio Futagawa. Paul Rudolph, Architectural Drawings. New York: Architectural Book Publishers, 1981. 5/ Paul Rudolph: Drawings for the Art and Architecture Building at Yale 1959 – 1963. New Haven: Yale University School of Architecture, 1988.

菲利普斯埃克赛特中学图书馆　PHILLIPS EXETER ACADEMY LIBRARY

1/ Wiggins, Glenn E. Louis I. Kahn: The Library at Phillips Exeter Academy. New York: Wiley & Sons, Incorporated, 1997.
2/ Futagawa, Yukio.Global Architecture 35: Kahn, Louis I. 1901 - 1974. Tokyo: A.D.A. EDITA, 1975.
3/ Gast, Klaus-Peter. Louis I Kahn: Complete Works. Munich: DVA, 2002.
4/ McCarter, Robert.Louis I Kahn.London: Phaidon, 2005.

AEG 透平机车间　AEG TURBINE FACTORY

1/ Buddensieg, Tilmann.Industriekultur: Peter Behrens and the AEG. Cambridge: The MIT Press, 1984. 2/ Anderson, Stanford. Peter Behrens and a New Architecture for the Twentieth Century. Cambridge: The MIT Press, 2000. 3/ Sennott, R. Stephen.Encyclopedia of 20th-Century Architecture. New York: Fitzroy Dearborn, 2004. 4/ Anderson, Stanford. "Modern Architecture and Industry: Peter Behrens and the AEG Factories." In Oppositions 23, Winter 1981. Cambridge: The MIT Press, 1981.

罗比住宅　FREDERICK C. ROBIE HOUSE

1/ Davis, Colin.Key Houses of the Twentieth Century. New York: W.W. Norton, 2006. 2/ Grehan, Farrell, and Frank Lloyd Wright.Visions of Wright. Boston: Bulfinch, 1997. 3/ Wright, Frank Lloyd.The Robie House. Palos Park, IL: Prairie School, 1968. 4/ Hoffmann, Donald, and Frank Lloyd Wright.Frank Lloyd Wright's Robie House: The Illustrated Story of an Architectural Masterpiece. New York: Dover Publishers, 1984.

申德勒自宅　SCHINDLER HOUSE

1/ Smith, Kathryn, and Grant Mudford.Schindler House. New York: Harry N. Abrams, 2001. 2/ Schindler, R. M.; Elizabeth A. T. Smith, and Michael Darling.The Architecture of R.M. Schindler. Los Angeles, CA: Museum of Contemporary Art, Los Angeles, 2001. 3/ Schindler, R. M., and Peter Noever.MAK Center for Art and Architecture: R .M. Schindler. Munich: Prestel, 1995. 4/ March, Lionel, and Judith Sheine.RM Schindler: Composition and Construction. London: Academy Editions, 1993. 5/ Gebhard, David.Schindler. Santa Barbara: Peregrine Smith, Inc., 1980. 6/ Sarnitz, August.R.M. Schindler, Architect: 1887 - 1953. New York: Rizzoli, 1988.

巴西议会大厦　NATIONAL CONGRESS, BRAZIL

1/ Macedo, Danilo Matoso.Congresso Nacional: Procedimentos projetuais e arquitetura brutalista. Curitiba: X do.co.mo.mo brasil, 2013. 2/ Carranza, Luis E., and Fernando Luiz Lara. Modern Architecture in Latin America: Art, Technology, and Utopia. Austin: University of Texas Press, 2015. 3/ Galiano, Luis Fernandez.Oscar Niemeyer: One Hundred Years. Madrid:

Arquitectura Viva, 2007. 4/ Mindlin, Henrique E. Modern Architecture in Brazil. Rio de Janeiro: Colibris Editora Ltda, 1956.

包豪斯德绍校舍　BAUHAUS DESSAU

1/ Irrgang, Christin.The Bauhaus Building in Dessau. Leipzig: Spector Books, 2014. 2/ Baumann, Kirsten.Bauhaus Dessau: Architecture—Design—Concept. Berlin: Jovis, 2007. 3/ Sharp, Dennis.Bauhaus, Dessau: Walter Gropius. New York: Phaidon, 2002. 4/ Colini, Laura, and Frank Eckardt, eds. Bauhaus and the City: A Contest[ed] Heritage for a Challenging Future. Würzburg: Königshausen & Neumann, 2011. 5/ Tafuri, Manfredo, and Francesco Dal Co. Modern Architecture, Vol. 1. Milan: Electa Editrice, 1976.

香港汇丰银行大厦　HONGKONG AND SHANGHAI BANK HEADQUARTERS

1/ Williams, Stephanie.Hong Kong Bank: The Building of Norman Foster's Masterpiece. London: Cape, 1989. 2/ Foster, Norman.Norman Foster: Drawings 1958‒2008. London: Ivory Press, 2010. 3/ Hongkong and Shanghai Bank Headquarters. (n.d.).Retrieved from http://www.fosterandpartners.com/projects/hongkong-and-shanghai-bank-headquarters/.4/ Sudjic, Deyan.New Architecture: Foster, Rogers, Stirling. London: Thames and Hudson Ltd., 1986.

格拉斯哥艺术学校　GLASGOW SCHOOL OF ART

1/ Grigg, Jocelyn.Charles Rennie Mackintosh. Salt Lake City: Gibbs Smith, 1988. 2/ Plunkett, Drew; Peter Trowles, Paul Heyer, and Shashi Caan.Four Studies on Charles Rennie Mackintosh. New York: New York School of Interior Design, 1996. 3/ Young, Andrew McLaren.Charles Rennie Mackintosh (1808‒1928): Architecture, Design and Painting. Edinburgh: Edinburgh Festival Society, 1968. 4/ Buchanan, William. Mackintosh's Masterwork: Charles Rennie Mackintosh and the Glasgow School of Art. San Francisco: Chronicle, 1989. 5/ Macaulay, James, and Mark Fiennes.Charles Rennie Mackintosh. New York: W.W. Norton, 2010.

玛丽亚别墅　VILLA MAIREA

1/ Aalto, Alvar.Villa Mairea. Helsinki: Alvar Aalto Foundation, 1998.
2/ Davis, Colin.Key Houses of the Twentieth Century. New York: W.W. Norton, 2006. 3/ Weston, Richard.Villa Mairea (Architecture in Detail). London: Phaidon Press, 2002. 4/ Gullichsen, Kirsi, and Ulla Kinnunen, eds. Inside the Villa Mairea: Art, Design, and Interior Architecture. Jyväskylä: Alvar Aalto Museum/Mairea Foundation, 2010. 5/ Blomstedt, Aulis; Alvar Aalto, and Aino Aalto.Villa Mairea. Jyväskylä: Alvar Aalto Museum, 1981. 6/ Aalto, Alvar, and Karl Fleig, eds. Alvar

Aalto: Band I 1922–1962. Zurich: Les Editions d'Architecture Artemis, 1963.

昌迪加尔议会大厦　THE ASSEMBLY, CHANDIGARH
1/ Boesiger, W., ed. Le Corbusier; OEuvre Complete 1957–1965. New York: George Wittenborn, Inc., 1965. 2/ Corbusier, Le.Creation Is a Patient Search. New York: Praeger, 1960. 3/ Högner, Bärbel; Clemens Kroll, Arthur Rüegg, Arno Lederer, and Mahendra Narain Sharma. Chandigarh: Living with Le Corbusier. Berlin: Jovis Verlag, 2010. 4/ Corbusier, Le.Chandigarh: City and Musée. New York: Garland Publishers, 1983. 5/ Scheidegger, Ernst; Maristella Casciato, and Stanislaus von Moos. Chandigarh 1956: Le Corbusier, Pierre Jeanneret, Jane B. Drew, E. Maxwell Fry. Zürich: Scheidegger & Spiess, 2010.

斯德哥尔摩公共图书馆　STOCKHOLM PUBLIC LIBRARY
1/ Adams, Nicholas.Gunnar Asplund. Milan: Mondadori Electa, 2011.
2/ Frampton, Kenneth, and Yukio Futagawa.Modern Architecture 1920–1945. New York: Rizzoli, 1983. 3/ Holmdahl, Gustav; Sven Ivar Lind, and Kjell Ödeen.Gunnar Asplund Architect, 1885–1940. Stockholm: AB Tidskriften Byggmästaren, 1950. 4/ Wrede, Stuart.The Architecture of Erik Gunnar Asplund. Cambridge: The MIT Press, 1980. 5/ Jones, Peter Blundell.Gunnar Asplund. New York: Phaidon Press Inc., 2006.

代代木国立综合体育馆　NATIONAL OLYMPIC GYMNASIUM
1/ Boyd, Robin.Kenzo Tange. New York, G. Braziller, 1962. 2/ Tange, Kenzo, and Udo Kultermann.Kenzo Tange, 1946–1969; Architecture and Urban Design. New York: Praeger Publishers, 1970. 3/ Von der Muhll, H.R; Kenzo Tange and Udo Kultermann. Kenzo Tange. Zurich: Verlag fur Architekur Artemis, 1978. 4/ Kuan, Seng; Yukio Lippit and Harvard University, GSD.Kenzo Tange: Architecture for the World. Baden: Lars Muller; London: Springer [distributor], 2012.

仙台媒体中心　SENDAI MEDIATHEQUE
1/ Gregory, Rob. Key Contemporary Buildings: Plans, Sections, and Elevations. New York: W.W. Norton & Company, 2008. 2/ Witte, Ron, ed. Toyo Ito, Sendai Mediatheque (CASE series). Munich: Prestel, 2002. 3/ Ito, Toyo, and Mutsuro Sasaki.Toyo Ito: Sendai Mediatheque. Miyagi, Japan; 1995–2000 (GA Detail 2). Tokyo: ADA EDITA, 2001. 4/ Ito, Toyo; Riken Yamamoto, Dana Buntrock, and Taro Igarashi.Toyo Ito. New York: Phaidon Press, 2009.5/ Maffei, Andrea, ed. Toyo Ito: Works, Projects, Writings. Milan: Electa, 2001.

爱因斯坦天文台　EINSTEIN TOWER
1/ Cobbers, Arnt.Erich Mendelsohn, 1887–1953: The Analytical

Visionary. Los Angeles: Taschen, 2007. 2/ Hentschel, Klaus. The Einstein Tower: An Intertexture of Dynamic Construction, Relativity Theory, and Astronomy. Stanford: Stanford University Press, 1997. 3/ Zevi, Bruno.Erich Mendelsohn: The Complete Works. Basel: Birkhäuser Publishers, 1999. 4/ Mendelsohn, Erich.Erich Mendelsohn: Architekt, 1887 - 1953: Gebaute Welten: Arbeiten für Europa, Palästina und Amerika. Ostfildern-Ruit: Hatje, 1998. 5/ Morgenthaler, Hans Rudolf.The Early Sketches of German Architect Erich Mendelsohn (1887 - 1953): No Compromise with Reality. Lewiston, NY: E. Mellen Press, 1992. 6/ Von Eckardt, Wolf.Eric Mendelsohn. New York: George Braziller, Inc., 1960.

利华大厦　LEVER HOUSE

1/ Danz, Ernst-Joachim.SOM: Architecture of Skidmore, Owings & Merrill, 1950 - 1962. New York: Monacelli, 2009. 2/ Bussel, Abby.SOM Evolutions: Recent Work of Skidmore, Owings & Merrill. Boston: Birkhäuser, 2000.
3/ Weston, Richard.Key Buildings of the 20th Century: Plans, Sections and Elevations. New York: W.W. Norton, 2010.

钻石农场中学　DIAMOND RANCH HIGH SCHOOL

1/ Kipnis, Jeffrey, and Todd Gannon, ed. Morphosis: Diamond Ranch High School, Diamond Bar, California; Thomas Blurock Architects, Executive Architects. New York: Monacelli Press, 2001. 2/ Gregory, Rob. Key Contemporary Buildings: Plans, Sections and Elevations. New York: W.W. Norton & Company, 2008. 3/ Croft, Catherine.Concrete Architecture. London: Lauren King Publishing Ltd., 2004. 4/ Hille, R. Thomas.Modern Schools: A Century of Design Education. Hoboken, New Jersey: John Wiley & Sons, Inc., 2011.5/ Futagawa, Yukio.GA Document Extra 09: Morphosis. Tokyo: A.D.A. EDITA, 1997.

玻璃屋　GLASS HOUSE

1/ Johnson, Philip.Johnson House, New Canaan, Connecticut, 1949 - edited by Yukio Futagawa, Tokyo: A.D.A. EDITA, 1972. 2/ Cassidy-Geiger, Maureen.The Philip Johnson Glass House: An Architect in the Garden. New York: Skira/Rizzoli, 2016. 3/ Heyer, Paul.American Architecture: Ideas and Ideologies in the Late Twentieth Century. New York: Van Nostrand Reinhold, 1993. 4/ Saunders, William S. Modern Architecture. New York: Harry N. Abrams, Publishers, 1990. 5/ Sharp, Dennis.Twentieth Century Architecture: A Visual History. New York: Facts on File, 1990. 6/ Nakamura, Toshio, ed. Glass House. New York: The Monacelli Press, 2007.

福特基金会总部大楼　FORD FOUNDATION HEADQUARTERS

1/ Futagawa, Yukio; Hiroshi Hara, Kevin Roche, John Dinkeloo and Associates.The Ford Foundation Building, New York, 1967; The Oakland Museum, California, 1969. Tokyo: A.D.A.

EDITA, 1971. 2/ Huxtable, Ada Louise. "Bold Plan for Building Revealed." The New York Times. 1967. 3/ Roche, Kevin, and Francesco Dal Co. Kevin Roche. New York: Rizzoli, 1985. 4/ Weston, Richard.Key Buildings of the Twentieth Century: Plans, Sections, and Elevations. New York: W.W. Norton, 2004. 5/ Hozumi, Nobuo.The Ford Foundation Headquarters, New York, N.Y., 1963 - 68. Tokyo: A.D.A. EDITA, 1977.

慕尼黑奥林匹克体育场　OLYMPIC STADIUM, MUNICH

1/ Glaeser, Ludwig.The Work of Frei Otto. Connecticut: MoMA, New York Graphic Society, 1972. 2/ Songel, Juan María. A Conversation with Frei Otto. New York: Princeton Architectural Press, 2010. 3/ Otto, Frei; Bodo Rasch, Gerd Pfafferodt, Adelheid Grafin Schonborn, and Sabine Schanz.Frei Otto, Bodo Rasch: Finding Form, Towards an Architecture of the Minimal. Stuttgart: Axel Menges, 1995. 4/ Meissner, Irene, and Eberhard Möller. Frei Otto: Forschen, Bauen, Inspirieren=Frei Otto-a Life of Research, Construction and Inspiration. München: Detail-Institut Für Internationale Architektur-Dokumentation, 2015. 5/ "Die Verwirklichung einer Idee. Anlagen u. Bauten für die Olympischen Spiele 1972." Bauen+Wohnen (special issue), 1972.

盖里自宅　GEHRY HOUSE

1/ Arnell, Peter; Ted Bickford, ed., and Mason Andrews. Frank Gehry, Buildings and Projects. New York: Rizzoli, 1985. 2/ Rappolt, Mark, and Robert Violette, ed. Gehry Draws. Cambridge: MIT Press, 2004. 3/ Gehry, Frank O.; Yukio Futagawa, Robert Violette.Frank O. Gehry: Gehry Residence. Tokyo: A.D.A. EDITA, 2015. 4/ Dal Co, Francesco, and Kurt W. Forester.Frank O. Gehry: The Complete Works. New York: The Monacelli Press, 1997.

伊瓜拉达墓园　IGUALADA CEMETERY

1/ Zabalbeascoa, Anatxu.Igualada Cemetery: Enric Miralles and Carme Pinós. London: Phaidon, 1996. 2/ Reed, Peter. Groundswell: Constructing the Contemporary Landscape. New York, NY: Museum of Modern Art, 2005.
3/ Berrizbeitia, Anita, and Linda Pollak.Inside Outside: Between Architecture and Landscape. Gloucester: Rockport Press, 1999. 4/ Betsky, Aaron.Landscrapers: Building with the Land. New York, NY: Thames & Hudson, 2002. 5/ Buchanan, Peter; Josep Maria Montaner, Dennis L. Dollens, and Lauren Kogod. The Architecture of Enric Miralles and Carme Pinós. New York: SITES/Lumen Books, 1990.

克朗楼　CROWN HALL

1/ Mies van der Rohe, Ludwig.The Mies van der Rohe Archive. New York: Garland Pub, 1986. 2/ Blaser, Werner.Mies Van Der Rohe — IIT Campus: Illinois Institute of Technology, Chicago. Basel: Birkhäuser, 2002. 3/ Mies van der Rohe, Ludwig; Yukio

Futagawa, and Ludwig Glaeser.Crown Hall, IIT, Chicago, Illinois, U.S.A., 1952－56: New National Gallery, Berlin, West Germany, 1968. Tokyo: A.D.A. EDITA, 1972. 4/ Blaser, Werner, and Ludwig Mies van der Rohe.Mies Van Der Rohe: IIT Campus, Illinois Institute of Technology, Chicago. Basel: Birkhäuser, 2002.

母亲住宅　VANNA VENTURI HOUSE

1/ Venturi, Robert.Complexity and Contradiction in Architecture. New York: The Museum of Modern Art, 1966. 2/ Schwartz, Frederic; Vincent Scully, Jr., and Robert Venturi.Mother's House: The Evolution of Vanna Venturi's House in Chestnut Hill. New York: Rizzoli, 1992. 3/ Futagawa, Yukio; Paul Goldberger, Venturi and Rauch.Venturi and Rauch: Vanna Venturi House, Chestnut Hill, Philadelphia, Pa., 1962; Peter Brant House, Greenwich, Conn., 1973; Carll Tucker III House, Westchester County, NY, 1975. Tokyo: A.D.A. EDITA, 1976.4/ Blackwood, Michael; Robert Venturi, Denise Scott Brown, Stephen Plumlee, Michael Blackwood Productions, Westdeutscher Rundfunk, British Broadcasting Corporation, Television Service.Robert Venturi and Denise Scott Brown. New York: Michael Blackwood Productions, 2006 Video. 5/ Carnicero, Iñaqui.Louis Kahn y Robert Venturi: Coincidencias del Gianicolo a Chestnut Hill. Madrid: Departamento de Proyectos Arquitectónicos, 2015.

洛弗尔健康之家　LOVELL HEALTH HOUSE

1/ Hines, Thomas.Richard Neutra and the Search for Modern Architecture. New York: Rizzoli, 2006. 2/ Lavin, Sylvia. Form Follows Libido: Architecture and Richard Neutra in a Psychoanalytic Culture. Cambridge: The MIT Press, 2004. 3/ Weston, Richard.Key Buildings of the 20th Century: Plans, Sections and Elevations. New York: W.W. Norton & Co., 2010. 4/ Davies, Colin.Key Houses of the Twentieth Century. New York: W.W. Norton, 2006.

图根哈特住宅　TUGENDHAT HOUSE

1/ Hammer-Tugendhat, Daniela; Ivo Hammer, and Wolf Tegethoff.Tugendhat House: Ludwig Mies van der Rohe. Vienna: Springer, 2015. 2/ Futagawa, Yoshio, and Yukio Futagawa.Mies van der Rohe: Villa Tugendhat Brno, Czechoslovakia,1928－30. Tokyo, Japan: A.D.A. EDITA, 2016. 3/ Mies van der Rohe, Ludwig; Beatriz Colomina, Moisés Puente, and Hans-Christian Schink. Mies van der Rohe: Casas. Barcelona: GG, 2009. 4/ Dunster, David.Key Buildings of the Twentieth Century. Vol. 1. London: Architectural Press, 1985.

贝克公寓　BAKER HOUSE

1/ Anderson, Stanford; Gail Fenske, and David Fixler.Aalto and America. New Haven: Yale University Press, 2012. 2/ Menin,

Sarah.Baker House: Aalto at M.I.T. Newcastle: Newcastle University, 1986. Dissertation.3/ Ray, Nicholas.Alvar Aalto. New Haven: Yale University Press, 2005. 4/ Schildt, Göran, and Alvar Aalto.Alvar Aalto in His Own Words. New York: Rizzoli, 1998. 5/ Hästesko, Arne, and Alvar Aalto.Alvar Aalto: What & When. Helsinki: Rakennustieto Publishing, 2014. 6/ Fleig, Karl, and Alvar Aalto, eds. Alvar Aalto; Band I 1922–1962. Zurich: Les Editions d'Architecture Artemis, 1963.

巴拉甘自宅与工作室　BARRAGÁN HOUSE AND STUDIO

1/ Alfaro, Alfonso; Daniel Garza Usabiaga, and Juan Palomar. Luis Barragán: His House. Mexico City: RM, 2011. 2/ Barragán, Luis; Yukio Futagawa, and Emilio Ambasz. House and Atelier for Luis Barragán, Tacubaya, Mexico City, 1947: Los Clubes, Suburb of Mexico City, 1963–69: San Cristobal, Suburb of Mexico City, 1967–68 (With the Collaboration of Arch. Andrés Casillas). Tokyo: A.D.A. EDITA, 1997. 3/ Futagawa, Yoshio, and Yukio Futagawa.Luis Barragán: Barragán House: Mexico City, Mexico 1947–48. Tokyo: A.D.A. EDITA Tokyo, 2009. 4/ Louis de Malave, Florita Z. Luis Barragan, the Architect and His Work. Monticello, Ill: Vance Bibliographies, 1983.

拜内克古籍善本图书馆　THE BEINECKE RARE BOOK & MANUSCRIPT LIBRARY

1/ Krinsky, Carol Herselle.Gordon Bunshaft of SOM. Cambridge: The MIT Press, 1988. 2/ Pavan, Vincenzo.Scriptures in Stone: Tectonic Language and Decorative Language. Milan: Skira, 2001. 3/ Perez, Adelyn. "AD Classics: Beinecke Rare Book and Manuscript Library/Gordon Bunshaft of Skidmore, Owings, & Merrill." ArchDaily.<http://www.archdaily.com/65987/ad-classics-beinecke-rare-book-and-manuscript-library-skidmore-owings-merrill>, 2010.

圣保罗艺术博物馆　MUSEU DE ARTE DE SÃO PAULO

1/ Oliveira, Olivia de.Lina Bo Bardi Built Work; 2G – Revista Internacional de Arquitectura (n° 23/24). Sao Paulo: Editora Gustavo Gilli, 2002. 2/ Frers, Lars, and Lars Meier.Encountering Urban Places, Visual and Material Performances in the City. Hampshire: Ashgate Publishing Ltd, 2007. 3/ Carranza, Luis E., and Fernando Luiz Lara.Modern Architecture in Latin America: Art, Technology, and Utopia. Austin: University of Texas Press, 2015.

布里翁墓园　BRION FAMILY TOMB

1/ McCarter, Robert.Carlo Scarpa. London: Phaidon Press, 2013. 2/ Dal Co, Francesco, and Giuseppe Mazzariol.Carlo Scarpa: The Complete Works. New York: Electa/Rizzoli, 1985. 3/ Beltramini, Guido, and Italo Zannier, eds. Carlo Scarpa: Architecture and Design. New York: Rizzoli, 2007. 4/ Saito, Yutaka, Hiroyuki Toyoda, and Nobuaki Furuya.Carlo Scarpa. Toyko: TOTO Shuppan, 1997.

孟加拉国国会大厦　NATIONAL ASSEMBLY BUILDING, BANGLADESH

1/ Kahn, Louis I.; Yukio Futagawa, and Kazi K. Ashraf.Louis I. Kahn, National Capital of Bangladesh, Dhaka, Bangladesh, 1962 - 83. Tokyo: A.D.A. EDITA Tokyo, 1994. 2/ Kahn, Louis I.; David B. Brownlee, Kate Norment, and David G. De Long. Louis I. Kahn: In the Realm of Architecture. New York: Rizzoli, 1991. 3/ Kries, Mateo; Jochen Eisenbrand, and Stanislaus von Moos. Louis Kahn: The Power of Architecture. Weil Am Rhein: Vitra Design Museum, 2012.4/ Kahn, Nathaniel, dir.; Louis Kahn Project, Inc.; Mediaworks, Inc.; HBO/Cinemax Documentary Films; New Yorker Films.My Architect: A Son's Journey. DVD, 2003.5/ McCarter, Robert.Louis I. Kahn. London: Phaidon, 2005.

斯图加特美术馆　NEUE STAATSGALERIE

1/ Rodiek, Thorsten.James Stirling Die Neue Staatsgalerie Stuttgart. Baden-Baden: Verlag Gerd Hatje, 1984. 2/ Baratelli, Alberto.James Stirling: La Galleria di Stato di Stoccarda. Florence: Alinea, 2002. 3/ Barthelmess, Stephan.Das postmoderne Museum als Erscheinungsform von Architektur. Munich: Tuduv, 1988. 4/ Maxwell, Robert.James Stirling: Writings on Architecture. London: Skira, 1998. 5/ Arnell, Peter; Ted Bickford, James Stirling, Michael Wilford and Associates. James Stirling: Buildings and Projects; James Stirling, Michael Wilford, and Associates. New York: Rizzoli, 1984.

西班牙国家罗马艺术博物馆　NATIONAL MUSEUM OF ROMAN ART

1/ Weston, Richard.Key Buildings of the 20th Century: Plans, Sections and Elevations. New York: W.W. Norton & Co., 2010. 2/ Jodidio, Philip.Architecture Now! Museums. Cologne: Taschen, 2010. 3/ Cortes, Juan Antonio.Rafael Moneo: International Portfolio 1985 - 2012. Germany: Axel Menges, 2013. 4/ Gonzalez, Francisco, and Nicholas Ray.Rafael Moneo: Building, Teaching, Writing. Connecticut: Yale University Press, 2015. 5/ Casamonti, Marco.Rafael Moneo (Minimum, Essential Architecture Library). Milan: Motta, 2009.

波尔多住宅　MAISON À BORDEAUX

1/ Koolhaas, Rem. OMA Rem Koolhaas Living, Vivre, Leben. Boston: Birkhäuser Verlag, 1998. 2/ Kara, Hanif; Andreas Georgoulias, and Jorge Silvetti.Interdisciplinary Design: New Lessons from Architecture and Engineering. Barcelona: Actar, 2012. 3/ Kolarevic, Branko, and Vera Parlac.Building Dynamics: Exploring Architecture of Change. New York: Routledge, 2015. 4/ Böck, Ingrid, and Rem Koolhaas. Six Canonical Projects by Rem Koolhaas: Essays on the History of Ideas. Berlin: Jovis, 2015.

米拉公寓　CASA MILÀ

1/ Güell, Xavier. Antoni Gaudí. Barcelona: Editorial Gustavo Gili, 1992.
2/ Zerbst, Rainer.Gaudí, 1852–1926: Antoni Gaudí i Cornet: A Life Devoted to Architecture. Cologne: Benedikt Taschen Verlag, 1988. 3/ Cuito, Aurora; Cristina Montes, and Antoni Gaudí. Antoni Gaudí: Complete Works. Cologne: Evergreen, 2009. 4/ Gaudí, Antoni; Michel Tapié, Joaquim Gomis, and Joan Prats Vallès. Gaudi—La Pedrera. Barcelona: Poliígrafa, 1971. 5/ Collins, George R. Antoni Gaudí. New York: George Braziller, 1960.

赛于奈察洛市政厅　SÄYNÄTSALO TOWN HALL

1/ Schildt, Göran, and Alvar Aalto. Alvar Aalto: Masterworks. New York: Universe Pub, 1998. 2/ Trencher, Michael.The Alvar Aalto Guide. New York: Princeton Architectural Press, 1996. 3/ Aalto, Alvar; Richard Weston, and Simo Rista.Town Hall, Säynaätsalo: Alvar Aalto. London: Phaidon Press, 1994. 4/ Muto, Akira; Yukio Futagawa, and Alvar Aalto.Alvar Aalto: Town Hall in Säynätsalo, Säynätsalo, Finland, 1950–52; Public Pensions Institute (Kansanelä kelaitos), Helsinki, Finland 1952–56. Tokyo: A.D.A. EDITA, 1981.

伦敦劳埃德大厦　LLOYD'S OF LONDON

1/ Rogers, Richard, & Architects.Richard Rogers + Architects: From the House to the City. London: Fiell, 2010. 2/ Cole, Barbie Campbell, and Ruth E. Rogers.Richard Rogers + Architects. New York: St. Martin's Press, 1985.
3/ Rogers, Richard, + Partners.Lloyd's of London. Milan: Edizioni Tecno, 1985. 4/ Powell, Ken, and Richard Partnership. Lloyd's Building: Richard Rogers Partnership. London: Phaidon Press, 1994. 5/ Gibbs, David.Building Lloyd's. London: Pentagram Design, 1986. 6/ Cook, Peter, and Richard Rogers.Richard Rogers + Partners: An Architectural Monograph. New York: St. Martin's Press, Inc., 1985.

阿拉伯世界研究中心　INSTITUT DU MONDE ARABE

1/ Montaner, Josep Maria.Neue Museen: Räume Für Kunst Und Kultur. Stuttgart: Krämer, 1990. 2/ Lacroix, Hugo.L'Institut Du Monde Arabe. Paris: La Différence, 2007. 3/ Noever, Peter, and Regina Haslinger.Architecture in Transition: Between Deconstruction and New Modernism. Munich: Prestel, 1991. 4/ Nouvel, Jean.Jean Nouvel, His Recent Works, 1987–1990. Barcelona: Colegio De Arquitectos De Cataluña, 1989. 5/ Boissiere, Olivier.Jean Nouvel: Jean Nouvel, Emmanuel Cattani and Associates. Zurich: Artemis Velags-AG, 1992.

拉维莱特公园　PARC DE LA VILLETTE

1/ Tschumi, Bernard; Jacques Derrida, and Anthony Vidler. Tschumi Parc de la Villette. London: Artifice, 2014. 2/ Tschumi,

Bernard.Cinégram Folie: Le Parc de La Villette. Princeton: Princeton Architectural Press, 1987. 3/ de Bure, Gilles; Jasmine Benjamin, and Lisa Palmer.Bernard Tschumi. Boston: Birkhäuser Verlag, 2008. 4/ Tschumi, Bernard.Architecture Concepts: Red Is Not a Color. New York: Rizzoli, 2012.

柏林犹太人博物馆　JEWISH MUSEUM BERLIN

1/ Libeskind, Daniel.Daniel Libeskind Jewish Museum Berlin. Barcelona: Ediciones Poligrafa, 2011. 2/ Libeskind, Daniel; Connie Wolf, Mitchell Schwarzer, and James Edward Young. Daniel Libeskind and the Contemporary Jewish Museum: New Jewish Architecture from Berlin to San Francisco. New York: Rizzoli, 2008. 3/ Schneider, Bernhard.Daniel Libeskind: Jewish Museum Berlin: Between the Lines. New York: Prestel, 1999. 4/ Binet, Hélène, and Raoul Bunschoten. A Passage Through Silence and Light. London: Black Dog Publishing, 1997.

代官山集合住宅　HILLSIDE TERRACE COMPLEX I - VI

1/ Maki, Fumihiko.Fumihiko Maki. London: Phaidon, 2009. 2/ Maki, Fumihiko, and Mark Mulligan.Nurturing Dreams: Collected Essays on Architecture and the City. Cambridge, MA: MIT, 2008. 3/ Taylor, Jennifer; Fumihiko Maki, and James Conner. The Architecture of Fumihiko Maki: Space, City, Order, and Making. Basel: Birkhäuser-Publishers for Architecture, 2003.4/ Maki, Fumihiko, and Arata Isozaki.New Public Architecture: Recent Projects by Fumihiko Maki and Arata Isozaki. New York: Japan Society, 1985. 5/ a+t research group (Aurora Fernández Per, Javier Mozas, Alex S. Ollero). "Slow City," in 10 Stories of Collective Housing, trans. Ken Mortimer, Vitoria-Gasteiz: a+t architecture publlishers, 2013.

史密斯住宅　SMITH HOUSE

1/ Meier, Richard.Richard Meier: Smith House, Darien, Connecticut, 1967 / House in Old Westbury, Long Island, New York, 1971. Tokyo: A.D.A. EDITA Tokyo, 1976. 2/ Clark, Roger H., and Michael Pause.Precedents in Architecture, 2E. New York: Van Nostrand Reinhold, 1996. 3/ Saunders, William S. Modern Architecture. New York: Harry N. Abrams Publishers, 1990. 4/ Eisenman, Peter; Michael Graves, Charles Gwathmey, John Hejduk, and Richard Meier.Five Architects: Eisenman, Graves, Gwathmey, Hejduk, Meier. London: Oxford University Press, 1975. 5/ Meier, Richard.Richard Meier, Architect, 1964 - 1984. New York: Rizzoli, 1984.

德国国家美术馆新馆　NEW NATIONAL GALLERY

1/ Mertins, Detlef, Ludwig Mies van der Rohe, and Phyllis Lambert.Mies. London: Phaidon, 2014. 2/ Vandenberg, Maritz, and Michael Brawne.Twentieth-Century Museums I: Ludwig Mies van der Rohe, New National Gallery, Berlin:Louis Kahn, Kimbell Art Museum: Richard Meier, Museum für

Kunsthandwerk. London: Phaidon, 1999. 3/ Vandenberg, Maritz, and Ludwig Mies van der Rohe.New National Gallery, Berlin, Ludwig Mies van der Rohe. London: Phaidon, 1998. 4/ Glaeser, Ludwig, and Yukio Futagawa.Mies van der Rohe: Crown Hall, IIT, Chicago, Illinois, U.S.A. 1952 - 56. New National Gallery, Berlin, West Germany, 1968. Tokyo: A.D.A. EDITA, 1974. 5/ Jarzombek, Mark. "Mies van der Rohe's New National Gallery and the Problem of Context," in Assemblage, No. 2. Cambridge: The MIT Press, 1987.

古堡博物馆　CASTELVECCHIO MUSEUM

1/ Murphy, Richard.Carlo Scarpa & Castelvecchio. Venice: Arsenale editrice, 1991. 2/ Marinelli, Sergio.Castelvecchio a Verona. Milan: Electa, 1991. 3/ Di Lieto, Alba.I disegni di Carlo Scarpa per Castelvecchio. Venice: Marsilio, 2006. 4/ Magnagnato, Licisco.Carlo Scarpa a Castelvecchio. Milan: Scotti, 1983.

莱萨浴场　PISCINA DE MARÉS

1/ Testa, Peter. Álvaro Siza. Boston: Birkhäuser Verlag, 1996. 2/ Moschini, Francesco. Álvaro Siza: l'architetto che voleva essere scultore. Galatina: Editrice Salentina, 2008. 3/ Siza, Álvaro, and Kenneth Frampton. Álvaro Siza: Complete Works. London: Phaidon, 2000. 4/ Jodidio, Philip; Álvaro Siza, Kristina Brigitta Köper, and Jacques Bosser. Álvaro Siza: Complete Works 1952 - 2013. Cologne: Taschen, 2013.

米歇尔广场公寓（路斯公寓）　MICHAELERPLATZ HOUSE (LOOSHAUS)

1/ Gravagnuolo, Benedetto.Adolf Loos. New York: Rizzoli, 1982. 2/ Tournikiotis, Panayotis.Adolf Loos. New York, NY: Princeton Architectural Press, 2002. 3/ Münz, Ludwig, and Gustave Künstler. Adolf Loos, Pioneer of Modern Architecture. New York: Praeger, 1966. 4/ Stewart, Janet.Fashioning Vienna: Adolf Loos's Cultural Criticism. London: Routledge, 2000. 5/ Pogačnik, Marko. Adolf Loos E Vienna: La Casa Sulla Michaelerplatz. Macerata: Quodlibet, 2011.

鲁萨科夫工人俱乐部　RUSAKOV WORKERS' CLUB

1/ Fosso, Mario, and Maurizio Meriggi.Konstantin S. Mel'nikov and the Construction of Moscow. Milan: Skira, 2001. 2/ Thiel-Siling, Sabine, and Wolfgang Bachmann.Icons of Architecture: The 20th Century. New York: Prestel, 1998. 3/ Starr, S. Frederick.Melnikov: Solo Architect in a Mass Society. Princeton: Princeton University Press, 1978. 4/ Wortmann, Arthur, and Konstantin Stepanovich Mel'nikov.Melnikov, The Muscles of Invention. Rotterdam: Van Hezik-Fonds 90, 1990.

棉纺织协会总部　MILL OWNERS' ASSOCIATION BUILDING

1/ Boesiger, W., ed. Le Corbusier et son atelier rue de

Sèvres 35: OEuvre complete 1952–1957. Zurich: Verlag für Architecktur, Artemis, 1957. 2/ Curtis, William J. R. Le Corbusier: Ideas and Forms. Oxford: Phaidon, 1986. 3/ Ching, Francis D. K.; Mark Jarzombek, and Vikramaditya Prakash.A Global History of Architecture. Hoboken: J. Wiley & Sons, 2011. 4/ Corbusier, Le; Yukio Futagawa, and Kenneth Frampton.Millowners Association Building, Ahmedabad, India, 1954: Carpenter Center for Visual Arts, Harvard University, Cambridge, Massachusetts, U.S.A. 1961–64. Tokyo, A.D.A. EDITA Tokyo, 1975.

斯塔尔住宅（22号案例住宅） STAHL HOUSE (CASE STUDY HOUSE NO. 22)

1/ Smith, Elizabeth A. T.; Julius Shulman, and Peter Gössel. Case Study Houses: The Complete CSH Program 1945–1966. Hong Kong: Taschen, 2009. 2/ Street-Porter, Tim.L.A. Modern. New York: Rizzoli, 2008. 3/ Jackson, N., and Peter Gössel. Pierre Koenig, 1925–2004: Living with Steel. Hong Kong: Taschen, 2007. 4/ Steele, James; Pierre Koenig, and David Jenkins.Pierre Koenig. London: Phaidon, 2002.

布罗伊尔楼（惠特尼美国艺术博物馆） THE BREUER BUILDING (WHITNEY MUSEUM OF AMERICAN ART)

1/ McCarter, Robert, and Marcel Breuer.Breuer, 2016. 2/ Bergdoll, Barry, and Massey, Jonathan.Marcel Breuer: Building Global Institutions. Zurich, Switzerland: Lars Muller Publishers, 2016. 3/ Cobbers, Arnt.Marcel Breuer: 1902–1981: Form Giver of the Twentieth Century. Köln: Taschen, 2007. 4/ Remmele, Mathias; Alexandra Pioch, Alexander von Vegesack, and Marcel Breuer.Marcel Breuer: Design and Architecture. Weil am Rhein: Vitra Design Museum, 2003.

屋顶加建 ROOFTOP REMODELING FALKESTRASSE

1/ Gössel, Peter, and Michael Mönninger. Coop Himmelb(l)au: Complete Works 1968–2010. Köln: Taschen, 2010. 2/ Offermann, Klaus.Architekten, Coop Himmelblau. Stuttgart: IRB Verlag, 1988. 3/ Prix, Wolf D.; Helmut Swiczinsky, Gudrun Hausegger, and Martina Kandeler-Fritsch.Coop Himmelblau Austria: From Cloud to Cloud: Biennale di Venezia 1996. Klagenfurt: Ritter, 1996. 4/ Vidler, Anthony.Warped Space: Art, Architecture, and Anxiety in Modern Culture. Cambridge: MIT, 2000. 5/ Moon, Karen.Modeling Messages: The Architect and the Model. New York: Monacelli, 2005.

拉金公司行政大楼 LARKIN COMPANY ADMINISTRATION BUILDING

1/ Quinan, Jack. Frank Lloyd Wright's Larkin Building: Myth and Fact. Cambridge: The MIT Press, 1987. 2/ Sáenz de Oiza, F.J., et al. AV Monografías 54: Frank Lloyd Wright. Madrid: Arquitectura Viva, 1995. 3/ Lind, Carla.Lost Wright: Frank Lloyd Wright's Vanished Masterpieces. New York: Simon & Schuster,

1996. 4/ Cattermole, Paul.Architectural Excellence: 500 Iconic Buildings. Richmond Hill: Firefly Books, 2008. 5/ Banham, Reyner. "The Services of the Larkin 'A' Building," The Journal of the Society of ArchitecturalHistorians, Vol. 37, No. 3 (Oct. 1978). Philadelphia: Society of Architectural Historians, 1978. Journal.6/ Scully, Vincent Jr. Frank Lloyd Wright. New York: George Braziller, Inc., 1960.

菲亚特工厂　FIAT WORKS

1/ Frampton, Kenneth.Modern Architecture 1851–1945. New York: Rizzoli International Publications, 1983. 2/ Sharp, Dennis. Twentieth Century Architecture: a Visual History. New York: Facts on File, 1990. 3/ Hofmann, Werner, and Udo Kultermann. Modern Architecture in Color. New York: The Viking Press, 1970. 4/ Kirk, Terry.The Architecture of Modern Italy, Volume II: Visions of Utopia, 1900–Present. New York: Princeton Architectural Press, 2005.

穆勒别墅　VILLA MÜLLER

1/ Loos, Adolf; Yehuda Safran, Wilfried Wang, and Mildred Budny.The Architecture of Adolf Loos: An Arts Council Exhibition. London: The Council, 1985. 2/ Colombian, Beatriz, "The Split Wall: Domestic Voyeurism," in Sexuality and Space. New York: Princeton University Press, 1992. 3/ Sarnitz, August. Adolf Loos, 1870–1933: Architect, Cultural Critic, Dandy. Cologne: Taschen, 2003. 4/ Tournikiotis, Panayotis.Adolf Loos. New York: Princeton Architectural Press, 2002. 5/ Loos, Adolf. Adolf Loos. Vienna: Graphische Sammlung Albertina, 1989. 6/ Risselada, Max, ed. Raumplan versus Plan Libre: Adolf Loos and Le Corbusier 1919–1930. New York: Rizzoli, 1988. 7/ Gravagnuolo, Benedetto.Adolf Loos. New York: Rizzoli, 1982.

范内勒工厂　VAN NELLE FACTORY

1/ Molenaar, Joris.Brinkman et Van der Vlugt Architects: Rotterdam's City-Ideal in International Style. Rotterdam: Nai010 Publ, 2012. 2/ Van der Oever, Martín. Van Nelle Fabriek in Stereo. Rotterdam: Uitgeverij 010 Publishers, 2002. 3/ J.A. Brinkman en L.C. Van der Vlugt (Firm); Jeroen Geurst, and Yukio Futagawa.J.A. Brinkman and L.C. van der Vlugt: Van Nelle Factory, Rotterdam, the Netherlands, 1925–31. Tokyo: A.D.A. EDITA Tokyo, 1994.

动态高效住宅　DYMAXION HOUSE

1/ Zung, Thomas T. K. Buckminster Fuller: Anthology for the Millennium. Carbondale: Southern Illinois University Press, 2014. 2/ Neder, Federico.Fuller Houses: R. Buckminster Fuller's Dymaxion Dwellings and Other Domestic Adventures. Baden, Switzerland: Lars Müller Publishers, 2008. 3/ Gorman, Michael John.Buckminster Fuller : Designing for Mobility. Milan: Skira, 2005. 4/ Hays, K. Michael, and Dana Miller.Buckminster Fuller:

Starting with the Universe. New York: Whitney Museum of American Art, in association with Yale University Press, 2008. 5/ Sieden, Lloyd Steven.Buckminster Fuller's Universe. New York: Plenum Press, 1989.

奎里尼·斯坦帕利亚基金会更新　QUERINI STAMPALIA RENOVATION

1/ McCarter, Robert.Carlo Scarpa. New York: Phaidon Press, 2013.
2/ Dal Co, Francesco.Carlo Scarpa: Fondazione Querini Stampalia a Venezia. Milan: Electa, 2006. 3/ Dal Co, Francesco, and Giuseppe Mazzariol.Carlo Scarpa: The Complete Works. New York: Electa/Rizzoli, 1985. 4/ Busetto, Giorgio.Cronaca Veneziana: Feste E Vita Quotidiana Nella Venezia Del Settecento: Vedute Di Gabriel Bella E Incisioni Di Gaetano Zompini Dalle Raccolte Della Fondazione Scientifica Querini Stampalia Di Venezia. Venice: Fondazione Scientifica Querini Stampalia, 1991.

六号住宅　HOUSE VI

1/ Eisenman, Peter and Antonini Saggio.Universale Di Architettura: Trivellazioni nel futuro. Torino: Testo & Immagine, 1996. 2/ Frank, Suzanne S., and Peter Eisenman.Peter Eisenman's House VI: The Client's Response. New York: Whitney Library of Design, 1994. 3/ Noever, Peter, and Peter Eisenman.Peter Eisenman: Barefoot on White-hot Walls. Ostfildern-Ruit: Hatje Cantz, 2005. 4/ Bradbury, Dominic, and Richard Powers.The Iconic House: Architectural Masterworks Since 1900. New York: Thames & Hudson, 2009. 5/ Luce, Kristina. "The Collision of Process and Form: Drawing's Imprint on Peter Eisenman's 'House VI.'" The Getty Research Journal, No. 2, 2010.6/ Eisenman, Peter; Rosalind Krauss, and Manfredo Tafuri.Houses of Cards: Critical Essays by Peter Eisenman, Rosalind Krauss, and Manfredo Tafuri. New York: Oxford University Press, 1987.

贝尔拉赫证券交易所　BEURS VAN BERLAGE

1/ Berlage, Hendrik Petrus.Hendrik Petrus Berlage: Disegni: IV Mostra Internazionale di Architettura. Venice: La Biennale di Venezia, 1986. 2/ Polano, Sergio, Hendrik Petrus Berlage, Giovanni Fanelli, Jan de Heer, and Vincent van Rossem.Hendrik Petrus Berlae. Milan: Electa Architecture, 2002. 3/ Berlage, Hendrik Petrus, and Iain Boyd Whyte.Hendrik Petrus Berlage: Thoughts on Style, 1886 – 1909. Santa Monica: Getty Center for the History of Art and the Humanities, 1996. 4/ Bock, Manfred; Jet Collee, Hester Coucke, and Maarten Kloos.Berlage in Amsterdam. Amsterdam: Architectura & Natura Press, 1992. 5/ Haags Gemeentemuseum.Berlage: Nederlandse Architectuur, 1856 – 1934. The Hague: Haags Gemeentemuseum, 1975. 6/ Tafuri, Manfredo, and Francesco Dal Co. Modern Architecture/1.

New York: Electa/Rizzoli, 1976. 7/ Zarzar, Karina Moraes. Innovation, Identity, and Sustainability in H.P. Berlage's Stock Exchange. Milan: XIII Generative Art Conference—Politecnico di Milano University, 2010.

甘布尔住宅　GAMBLE HOUSE

1/ Bosley, Edward R. Greene & Greene. London: Phaidon, 2003. 2/ Bosley, Edward.Gamble House: Greene & Greene. New York: Phaidon, 2002. 3/ Mackintosh, Charles Rennie; Charles F. A. Voysey, and James Macaulay.Arts & Crafts Houses 2, 2. London: Phaidon, 1999. 4/ Smith, Bruce, and AlexanderVertikoff.Greene and Greene: Master Builders of the American Arts and Crafts Movement. London: Thames and Hudson, 1998. 5/ Futagawa, Yukio, and Randell L. Makinson. Greene & Greene: David B. Gamble House, Pasadena, California, 1908. Tokyo: A.D.A. EDITA, 1984. 6/ Makinson, Randell L. Greene & Greene. Salt Lake City: Peregrine Smith, 1977.

洛弗尔海滨住宅　LOVELL BEACH HOUSE

1/ Steele, James, and Peter Gössel. R.M. Schindler 1887–1953: An Exploration of Space. Köln: Taschen, 2005.2/ Frampton, Kenneth, and Larkin, David.American Masterworks: The Twentieth Century House. New York: Rizzoli, 1995. 3/ Sarnitz, August, E. "Proportion and Beauty: The Lovell Beach House by Rudolph Michael Schindler, Newport Beach, 1922–1926". Journal of the Society of Architectural Historians/Society of Architectural Historians.374–388.

都林展览馆　TURIN EXHIBITION HALL

1/ Huxtable, Ada Louise.Pier Luigi Nervi. New York: George Braziller, Inc., 1960. 2/ Powell, Ken. The Great Builders. London: Thames & Hudson, 2011.3/ Pace, Sergio.Pier Luigi Nervi: Torino, La Committenza Industriale, Le Culture Architettoniche E Politecniche Italiane. Cinisello Balsamo: Silvana, 2011.4/ Nervi, Pier Luigi; Juergen Joedicke, and Ernst Priefert.The Works of Pier Luigi Nervi. London: Architectural Press, 1957. 5/ Sharp, Dennis.The Illustrated Encyclopedia of Architects and Architecture. New York: Whitney Library of Design, 1991.

阿姆斯特丹市孤儿院　MUNICIPAL ORPHANAGE, AMSTERDAM

1/ Strauven, Francis; Aldo van Eyck, and Herman Hertzberger. Aldo Van Eyck's Orphanage: A Modern Monument. Rotterdam: NAi, 1996. 2/ Hertzberger, Herman.Space and the Architect: Lessons in Architecture 2. Rotterdam: 010 Publishers, 2000. 3/ Ford, Edward R.The Architectural Detail. New York: Princeton Architectural Press, 2011. 4/ Simitch, Andrea, and Val Warke. The Language of Architecture: 26 Principles Every Architect Should Know. Beverly: Rockport Publishers, 2014. 5/ Ligtelijn, Vincent, ed. Aldo van Eyck: Works. Boston: Birkhäuser Publishers, 1999.

卡朋特视觉艺术中心　CARPENTER CENTER FOR THE VISUAL ARTS

1/ Boesiger, W., ed. Le Corbusier et son atelier rue de Sèvres 35: OEuvre complete 1957–1965. Zurich: Verlag für Architecktur, Artemis, 1965. 2/ Sekler, Eduard F., and William Curtis. Le Corbusier Work. Cambridge: Harvard University Press, 1978. 3/ Corbusier, Le; Yukio Futagawa, and Kenneth Frampton. Millowners Association Building, Ahmedabad, India, 1954: Carpenter Center for Visual Arts, Harvard University, Cambridge, Massachusetts, U.S.A. 1961–64. Tokyo: A.D.A. EDITA Tokyo, 1975. 4/ Corbusier, Le, and Richard Joseph Ingersoll. Le Corbusier: A Marriage of Contours. New York: Princeton Architectural Press, 1990. 5/ Brooks, Allen. Le Corbusier: Carpenter Center, Unité d'Habitation, Firminy, and Other Buildings and Projects, 1961–1963. New York: Garland Publication, 1984.

柏林自由大学　FREE UNIVERSITY OF BERLIN

1/ Feld, Gabriel, and Mohsen Mostafavi. Free University, Berlin. London: Architectural Association Publications, 1999. 2/ Avermaete, Tom. Another Modern: The Post-War Architecture and Urbanism of Candilis-Josic-Woods. Rotterdam: NAi Publishers, 2005. 3/ Krunic, Dina. The Groundscraper: Candilis-Josic-Woods' Free University Building, Berlin 1963–1973. Thesis. 2012. 4/ Smithson, Alison, ed. Team 10 Primer. Cambridge: The MIT Press, 1968.

鲍斯韦教堂　BAGSVÆRD CHURCH

1/ Norberg-Schulz, Christian, and Yukio Futagawa. Jörn Utzon, Church at Bagsvœrd, near Copenhagen, Denmark, 1973–76. Tokyo: A.D.A. EDITA Tokyo, 1981. 2/ Utzon, Jørn; Kjeld Kjeldsen, Michael Juul Holm, and Mette Marcus. Jørn Utzon: The Architect's Universe. Humlebœk, Denmark: Louisiana Museum of Modern Art, 2008. 3/ Møller, Henrik Sten, and Vibe Udsen. Jørn Utzon Houses. Copenhagen: Living Architecture Publishing, 2006. 4/ Weston, Richard. Key Buildings of the 20th Century: Plans, Sections and Elevations. New York: W.W. Norton, 2010.

庞培娅艺术中心　SESC POMPÉIA

1/ Oliveira, Olivia De. The Architecture of Lina Bo Bardi: Subtle Substances. Barcelona: Editorial Gustavo Gili, 2006. 2/ Oliveira, Olivia de. Lina Bo Bardi Built Work; 2G – Revista Internacional de Arquitectura (n° 23/24). São Paulo: Editora Gustavo Gilli, 2002. 3/ Veikos, Cathrine. Lina Bo Bardi: The Theory of Architectural Practice. New York: Routledge, 2014.

柯布西耶中心（海蒂·韦伯住宅）　CENTRE LE CORBUSIER (HEIDI WEBER HOUSE)

1/ Curtis, William J. R. Le Corbusier: Ideas and Forms. London:

Phaidon, 2015. 2/ Dumont d'Ayot, Catherine, and Tim Benton.Le Corbusier's Pavilion for Zurich: Model and Prototype of an Ideal Exhibition Space. Zürich: Müller, 2013. 3/ Heidi Weber: 50 Years Ambassador for Le Corbusier 1958–2008. Zurich: Birkhäuser book, 2010. 4/ Žaknić, Ivan. Le Corbusier Pavillion Suisse: The Biography of a Building. Basel: Birkhauser, 2004.

加拉拉特西公寓　GALLARATESE II APARTMENTS

1/ Frampton, Kenneth, and Vittorio Magnago Lampugnani. World Architecture 1900–2000 a Critical Mosaic.Volume 4. Wien: Springer, 1999.2/ Nicolin, Pierluigi, and Y. Futagawa.Carlo Aymonino/Aldo Rossi: Housing Complex at the Gallaratese Quarter, Milan, Italy. 1969–1974. Tokyo: A.D.A. EDITA, 1981. 3/ Rossi, Aldo, and Peter Eisenman.The Architecture of the City. Cambridge, Mass: MIT Press, 1984.

中央贝赫保险公司大楼　CENTRAAL BEHEER BUILDING

1/ Frampton, Kenneth.A Genealogy of Modern Architecture: Comparative Critical Analysis of Built Form. Zürich: Lars Müller Publishers, 2015. 2/ Weston, Richard.Key Buildings of the Twentieth Century: Plans, Sections and Elevations. New York: W.W. Norton & Co., 2010. 3/ Hertzberger, Herman.Architecture and Structuralism: The Ordering of Space. Amsterdam: NAi010 Publishers, 2015. 4/ Hertzberger, Herman.Cultuur onder dak. Rotterdam: Uitgeverij 010 Publishers, 2004.

维特拉消防站　VITRA FIRE STATION

1/ Futagawa, Yukio.GA Document Extra 03: Zaha Hadid. Tokyo: A.D.A. EDITA, 1996. 2/ Hadid, Zaha; Alexandra Papadakis, and A. Papadakes.Zaha Hadid: Testing the Boundaries. London: Papadakis Publisher, 2005. 3/ Jodidio, Philip. Hadid: Zaha Hadid Complete Works 1979–2013. Cologne: Taschen, 2013. 4/ Sisson, Patrick. "21 First Drafts: Zaha Hadid's Vitra Fire Station." Curbed.<http://www.curbed.com/2015/8/5/9933580/21-first-drafts-zaha-hadids-vitra-fire-station> 05 Aug 2015.

横滨国际港客运中心　YOKOHAMA INTERNATIONAL PORT TERMINAL

1/ Gregory, Rob. Key Contemporary Buildings: Plans, Sections, and Elevations. New York: W.W. Norton & Company, 2008. 2/ Ferre, Albert, Tomoko Sakamoto, Michael Kubo.The Yokohama Project: Foreign Office Architects. Barcelona: Actar, 2002. 3/ Carpo, Mario.The Digital Turn in Architecture, 1992–2012. Chichester: Wiley, 2013. 4/ Broto, Carles; Jacobo Krauel, Jay Noden, and William George.Transportation Facilities. Barcelona: LinksBooks, 2012.

图片来源与英文版参与人员

1/ Villa Savoye, Poissy, 1928 Paul Kozlowski ©2017 FLC/Artists Rights Society (ARS), New York. 2/ Chapelle Notre Dame du Haut, Ronchamp, 1950-1955 Photo : Paul Kozlowski ©2017 FLC/Artists Rights Society (ARS), New York. 3/ Barcelona Pavilion by Richard Moross (www.flickr.com/photos/richardmoross/4164341602/) CC BY 2.0. Modified. 4/ ©Katsuhisa Kida/FOTOTECA. 5/ Jack E. Boucher/Courtesy Library of Congress Prints & Photographs Division, HABS IS,51-RACI,5—1. 6/ Jack E. Boucher/Courtesy Library of Congress Prints & Photographs Division, ILL,47-PLAN. V,1-9. 7/ Louis Kahn's Salk Institute by Jason Taellious (https://www.flickr.com/photos/dreamsjung/3040455466/) CC BY 2.0. Modified. 8/ ©Mark Lyon. 9/ Couvent Sainte-Marie de la Tourette, Eveux-sur-l'Arbresle, 1953 Photo: Olivier Martin-Gambier ©2017 FLC/Artists Rights Society (ARS), New York. 10/ TWA Flight Center by Jim.henderson (https://commons.wikimedia.org/wiki/File:AirTrain_JFK_passes_TWA_Flt_Ctr_jeh.JPG) CC BY-SA 4.0 International. Desaturated, Cropped. 11/ Philharmonie by Matthais Rosenkranz (https://www.flickr.com/photos/rosenkranz/294605986/) CC BY 3.0. Desaturated. 12/ Guggenheim 23 by Tony Hisgett (https://www.flickr.com/photos/hisgett/3798502025/) CC BY 2.0. Desaturated, cropped. 13/ Jack E. Boucher/Courtesy Library of Congress Prints & Photographs Division HABS PA,26-OHPY.V,1--3. 14/ Skogskyrkogården - Enskede - Stockholm by Esther Westerveld (https://www.flickr.com/photos/westher/14943642596/) CC BY 2.0 / Desaturated, Cropped. 15/ Seagram Building by Jules Antonio (https://www.flickr.com/photos/julesantonio/6268091900/) CC BY-SA 2.0. 16/ The Solomon R. Guggenheim Museum by Jules Antonio (https://www.flickr.com/photos/julesantonio/11990550696/) CC BY-SA 2.0. Desaturated. 17/ Eames House = Mecca by Lauren Manning (https://www.flickr.com/photos/laurenmanning/8008065034/) CC BY 2.0. Modified. 18/ Leicester University Engineering Building by NotFromUtrecht (https://upload.wikimedia.org/wikipedia/commons/a/ad/Leicester_University_Engineering_Building.jpg) CC BY-SA 3.0. Desaturated, Cropped. 19/ Casa del Fascio, Architect Guiseppe Terragni. Permission granted by Archivio Terragni. 20/ Unité d'Habitation, Marseille, 1945 Photo : Paul Kozlowski 1997 ©2017 FLC/Artists Rights Society (ARS), New York. 21/ ©Timothy Hursley. 22/ Library of Congress, Prints & Photographs Division, photograph by Carol M. Highsmith LC-DIG-highsm-13209. 23/ Rietveld Schröderhuis winter 2014-15 02 by Luis Guillermo R. (https://commons.wikimedia.org/wiki/File%3ARietveld_Schr%C3%B6derhuis_winter_2014-15_02.JPG) CC BY 4.0 International. Desaturated, Cropped. 24/ Photo: Tomio Ohashi ©KISHO KUROKAWA architect & associates. 25/ Otto Wagner Postsparkasse Vienna - Dec 2014 - 2 by Andrew Nash (https://www.flickr.com/photos/andynash/16193628732/) CC BY-SA 3.0. Desaturated, Cropped. 26/ ©Eredi Curzio Malaparte. Image: Villa

Malaparte by Francois Phillip (https://www.flickr.com/photos/frans16611/4729750386/) used under CC-BY 2.0. Modified.27/ Public Domain.28/ Yale Art and Architecture Building by Gunnar Klack (https://commons.wikimedia.org/wiki/File:Yale-Art-and-Architecture-Building-Rudolph-Hall-New-Haven-Connecticut-Apr-2014.jpg) CC 4.0 BY SA.Modified.29/ Phillips Exeter Library, New Hampshire – Louis I. Kahn (1972) by Pablo Sanchez (https://www.flickr.com/photos/pablosanchez/3503858665/) CC BY 2.0.Modified.30/ DSC06900 by IK's World Trip (https://www.flickr.com/photos/ikkoskinen/2618814478/) CC BY 2.0. Modified.31/ Robie House, Hyde Park, by Frank Lloyd Wright by Naotake Murayama (https://www.flickr.com/photos/naotakem/9618352872/) CC BY 2.0.Modified.32/ ©MAK Center/Joshua White.33/ National Congress--Brasilia by Razvan Orendovici (https://www.flickr.com/photos/razvanorendovici/14630598183/) CC BY 2.0. Modified.34/ Dessau-Bauhaus by Spyrosdrakopoulos (https://commons.wikimedia.org/wiki/File:6251_Dessau.JPG) CC BY-SA 4.0. Desaturated, Cropped. 35/ ©Ian Lambot / Arcaid.36/ 080320101938 by IK's World Trip (https://www.flickr.com/photos/ikkoskinen/4418148288) CC BY 2.0. Modified.37/ 4Y1A7841 by Ninara (https://www.flickr.com/photos/ninara/26710745140/) CC BY-SA 2.0.Modified.38/ KIF_4646_Pano by duncid (https://www.flickr.com/photos/48013827@N00/275987220) CC BY-SA 2.0.Modified.39/ 810_5725 by Bengt Nyman (https://www.flickr.com/photos/97469566@N00/16649751719) CC BY-SA 2.0.Modified.40/ Yoyogi National Gymnasium by kanegen (https://www.flickr.com/photos/kanegen/3076874395) CC BY 2.0. Modified.41/ Photo provided by Miyagi Prefecture Sightseeing Section. 42/ La tour Einstein (Potsdam, Allemagne) by Jean-Pierre Dalbéra (https://www.flickr.com/photos/dalbera/9616566364) CC BY 2.0. Modified.43/ Photo: Ezra Stoller ©ESTO. 44/ ©Timothy Hursley.45/ Library of Congress, Prints & Photographs Division, photograph by Carol M. Highsmith LC-DIG-highsm-04817. 46/ ©Richard Anderson 47/ Blick vom Olympiaberg auf das Olympiastadion by Amrei-Marie (https://commons.wikimedia.org/wiki/File:Blick_vom_Olympiaberg_auf_das_Olympiastadion.jpg) CC BY-SA 4.0 International. Desaturated, cropped. 48/ Gehry House, I by IK's World Trip (https://www.flickr.com/photos/ikkoskinen/350055881) CC BY 2.0. Modified.49/ ©David Cabrera.50/ S. R. Crown Hall by Arturo Duarte Jr. (https://commons.wikimedia.org/wiki/File:S.R._Crown_Hall.jpg) CC BY-SA 3.0.Desaturated.51/ Photographer Unknown, Courtesy of Venturi, Scott Brown and Associates, Inc. 52/ Lovell House, Richard Neutra, Architect 1929 by MichaelJLocke (https://commons.wikimedia.org/wiki/File:Lovell_House,_Richard_Neutra,_Architect_1929.jpg) CC BY-SA 4.0. Desaturated, Cropped. 53/ Vila Tugendhat by Petr1987 (https://commons.wikimedia.org/wiki/File:Vila_Tugendhat_exterior_Dvorak2.JPG) CC BY-SA 4.0.Desaturated.54/ Public Domain.55/ Luis Barragán House and Studio Street view by Thomas Ledl (https://

en.wikipedia.org/wiki/File:Luis_Barrag%C3%A1n_House_and_Studio_Street_view.JPG) CC BY-SA 4.0 Internation. Desaturated, cropped. 56/ Photo: Ezra Stoller ©ESTO. 57/ Museu de Arte de Sao Paulo by Dave Lonsdale (https://www.flickr.com/photos/davelonsdale/5819164997) CC BY 2.0. Modified.58/ MAXXI Museo nazionale delle arti del XXI secolo, Roma. MAXXI Architettura Collection 59.Originali Arch. Scarpa [Tomba monumentale Brion], San Vito d'Altivole (TV).59/ Panorama_of_National_Assembly_of_Bangladesh by Nahid Sultan (https://commons.wikimedia.org/wiki/File:Panorama_of_National_Assembly_of_Bangladesh.jpg) CC BY-SA 3.0. Desaturated, cropped. 60/ Die neue Staatsgalerie, Stuttgart - James Stirling, 1982 by Timothy Brown (https://www.flickr.com/photos/atelier_flir/2793727265/) CC BY 2.0. Modified.61/ ©Michael Moran / OTTO.62/ By Hans Werlemann / Copyright OMA.63/ Casa Mila by Clarence (https://www.flickr.com/photos/bracketing_life/4104719680) CC BY 2.0. Modified.64/ SaynatsaloTown Hall by Zache (https://commons.wikimedia.org/wiki/File:SaynatsaloTownHall4.jpg) CC BY SA 3.0. Desaturated, Cropped. 65/ ©Richard Bryant / Arcaid.66/ ©2017 Artists Rights Society (ARS), New York / ADAGP, Paris.67/ ©Bernard Tschumi Architects.68/ ©Guenter Schneider.69/ Hillside_Terrace_A_B by Wiiii (https://upload.wikimedia.org/wikipedia/commons/8/89/Hillside_Terrace_A_B_2010.jpg) CC 3.0 BY SA.Modified.70/ ©Scott Frances/ OTTO.71/ Neue Nationalgalerie by Rosmarie Voegtli (https://www.flickr.com/photos/rvoegtli/3353185093) CC BY 2.0. Modified.72/ MAXXI Museo nazionale delle arti del XXI secolo, Roma. MAXXI Architettura Collection 98.[Museo di Castelvecchio], Verona.73/ Swimming Pool Piscinas de Maré s Leça da Palmeira by Christian Gänshirt (https://commons.wikimedia.org/wiki/File:Swimming_Pool_Piscinas_de_Mar%C3%A9s_Le%C3%A7a_da_Palmeira_by_%C3%81lvaro_Siza_foto_Christian_G%C3%A4nshirt.jpg) CC BY-SA 4.0. Desaturated, Cropped. 74/ P7099953_ShiftN by Paul Barker Hemings (https://www.flickr.com/photos/pollobarca/15276277612) CC BY-SA 2.0.Modified.75/ Rusakov Club, Moscow by NVO (https://commons.wikimedia.org/wiki/File:Moscow,_Stromynka_6_July_2009_06.JPG) CC BY-SA 3.0. Desaturated, Cropped. 76/ Mill Owners' Association Building, Ahmedabad by Sanyam Bahga (https://commons.wikimedia.org/wiki/File:ATMA_House_187.jpg) CC BY-SA 3.0. Desaturated, Cropped. 77/ Julius Schulman ©J. Paul Getty Trust. Getty Research Institute, Los Angeles (2004.R.10). 78/ Whitney Museum of American Art by Jonathan Lin (https://www.flickr.com/photos/jonolist/14857063277/) CC BY-SA 2.0.Modified.79/ ©Duccio Malagamba courtesy of COOP HIMMELB(L)AU. 80/ Courtesy of the Buffalo History Museum.81/ Courtesy of Centro Storico Fiat.82/ Villa Müller in Prag by Hpschaefer (https://commons.wikimedia.org/wiki/File:Villa-Mueller-Prag-2.) CC BY 3.0. Desaturated, Cropped. 83/ Van Nellefabriek by Hanno Lans (https://commons.wikimedia.org/wiki/File:Van_Nellefabriek_(20288847352).jpg) CC BY-SA 2.0. Modified.84/

Courtesy, The Estate of R. Buckminster Fuller. 85/ MAXXI Museo nazionale delle arti del XXI secolo, Roma. MAXXI Architettura Collection 111.[Sede delle Fondazione Querini Stampalia], Venezia.86/ Photo by Richard Frank, courtesy Eisenman Architects.87/ Beurs van Berlage: Overzicht Damrakzijde by G.J. Dukker (https://commons.wikimedia.org/wiki/File:Overzicht_Damrakzijde_-_Amsterdam_-_20286196_-_RCE.jpg) CC BY-SA 4.0 International. Cropped.88/ IMG_1238 Gamble House by Robert B. Moffatt (https://www.flickr.com/photos/55050575@N06/22392915088/) CC BY-SA 2.0.Modified.89/ Lovell_Beach_House_photo_D_Ramey_Logan (https://commons.wikimedia.org/wiki/File:Lovell_Beach_House_photo_D_Ramey_Logan.jpg) CC BY-SA 3.0.Desaturated.90/ Servizio fotografico (Torino, 1961) by Paolo Monti (https://commons.wikimedia.org/wiki/File:Paolo_Monti_-_Servizio_fotografico_(Torino,_1961)_-_BEIC_6337387.jpg) CC BY-SA 4.0 International. Courtesy Fondo Paolo Monti, owned by BEIC and located in the Civico Archivio Fotografico of Milan. 91/ Credit: Noelle Tay Li-Zhen. 92/ ©Scott Norsworthy.93/ Freie Universitaet Berlin - Gebaeudekomplex Rost- und Silberlaube by torinberl (https://commons.wikimedia.org/wiki/File:Freie_Universitaet_Berlin_-_Gebaeudekomplex_Rost-_und_Silberlaube.jpg) CC BY-SA 3.0 Unported.Desaturated.94/ jørn utzon, bagsværd kirke - bagsvaerd church, copenhagen 1967-1976 by seier+seier (https://www.flickr.com/photos/seier/5893565856) CC BY 2.0. Modified.95/ ©Pedro Kok.96/ Heidi Weber Pavilion by Le Corbusier by Fatlum Haliti (https://www.flickr.com/photos/lumlumi/439657731) CC BY 2.0. Modified.97/ Burçin Yildirim ©Eredi Aldo Rossi / Fondazione Aldo Rossi.98/ Hertzberger Centraal Beheer by Apdency (https://commons.wikimedia.org/wiki/File:Hertzberger_Centraal_Beheer1.jpg) CC BY-SA 3.0.Desaturated, Cropped. 99/ Feuerwehrhaus von Zaha Hadid bei Vitra in Weil am Rhein, links Schaudepot von Herzog & de Meuron by Andreas Schwarzkopf(https://commons.wikimedia.org/wiki/File:Feuerwehrhaus_von_Zaha_Hadid_bei_Vitra_in_Weil_am_Rhein,_links_Schaudepot_von_Herzog_%26_de_Meuron_1.jpg) CC BY-SA 3.0 Unported. Desaturated.100/ Credit: Satoru Mishima, courtesy of FMA.

The best efforts of the authors and the publisher have been employed in obtaining permissions for, and determining proper attribution of, the photographs shown in this book. If, by regretful oversight, any credits are incomplete or missing, the publisher will take the necessary steps to ameliorate this condition in a reprint, upon being made aware of any such oversight.

本书的作者与英文版出版商（RIZZOLI出版集团）已经尽了最大努力去确认书中照片的来源并获取使用许可，如仍有疏漏，出版商将在本书往后重印时尽力改善并弥补。

Project Manager（项目经理）
Eui-Sung Yi

Production Manager（生产经理）
Ryan Doyle

Book Design（书籍设计）
Lily Bakhshi

Content Management（内容管理）
Nicole Meyer

Sarah Moseley

Writing（撰文）
Val Warke

with

Eric Keune

Andrea Simitch

Foreword（前言）
Thom Mayne

Research and Production Assistants（研究助理）
Yitao Chen

Stanley Cho

Patrick Geske

Adeline Morin

Cameron Northrop

Beyza Paksoy

John Paul Salcido

Kevin Sherrod

Young Sun

Jane Suthi

Way Tang

Andrea Tzvetkov

译后记

　　普利兹克奖得主汤姆·梅恩与加州大学洛杉矶分校的同事筹措五年，终于编成了此书，将诸多不可能变成现实。汤姆·梅恩举 The Now Institute 师生之力，详细绘制 100 所建筑的平、立、剖面图与轴测图，反复甄别图形正误，做到了"正确"；他广泛邀请建筑界的朋友和同事，共有 50 位当代著名建筑师应邀为本书选择案例（其中普利兹克奖得主就有 12 位），从本行业最具专业性的人士手中筛选出 100 例现代建筑，做到了"客观"。如此费力出书的原因只是因为他觉得学生看得经典实例太少，想要为学生们做一本好的建筑参考书，其教师的责任感令人肃然起敬。

　　一般而言，唱片或书籍的"精选集"往往是将作者不同时期最受欢迎的作品拼合出来，虽然可以代表创作历程的大体风貌，却又因过杂过多而失去立场。本书中涉及的建筑实例时间跨度大、形式差异明显，作者却以统一的视角与分析方法对它们做出了极精练的解析，内容充实而有效，对于建筑初学者而言是一本不可多得的好书。本人在建筑教学工作中，每次看到学生用手机搜索各式各样的"范例"都觉得无可奈何，学生也因为找不到恰当的学习资料而挠头。希望这本书能够带领初学者认识现代建筑，开始更加完整、高效的案例学习之路。

　　本书的内容精练，代表了现代建筑中极少的一些作品，无法支撑深度学习研究。我认为，应该将本书看作一篇现代建筑作品的学习目录，用来指导学习和资料收集的目标与方向。书中"建筑师的选择"一章很有价值，作者列出了所有建筑师选择的作品案例，可以据此举一反三，引申出一份更为广泛的学习目标单。由于译者水平有限，翻译中的失误还望广大读者指正。本书的前言部分、"建筑 100 中"的前 50 例由张涵翻译，其余部分由樊敏翻译。

译者，2019 年初